ATELIER: ETHNOGRAPHIC INQUIRY
IN THE TWENTY-FIRST CENTURY

Kevin Lewis O'Neill, Series Editor

1. *Mortal Doubt: Transnational Gangs and Social Order in Guatemala City*, by Anthony W. Fontes
2. *Contingent Kinship: The Flows and Futures of Adoption in the United States*, by Kathryn A. Mariner
3. *Captured at Sea: Piracy and Protection in the Indian Ocean*, by Jatin Dua
4. *Fires of Gold: Law, Spirit, and Sacrificial Labor in Ghana*, by Lauren Coyle Rosen
5. *Tasting Qualities: The Past and Future of Tea*, by Sarah Besky
6. *Waste Worlds: Inhabiting Kampala's Infrastructures of Disposability*, by Jacob Doherty
7. *The Industrial Ephemeral: Labor and Love in Indian Architecture and Construction*, by Namita Vijay Dharia
8. *Pinelandia: An Anthropology and Field Poetics of War and Empire*, by Nomi Stone
9. *Stuck Moving: Or, How I Learned to Love (and Lament) Anthropology*, by Peter Benson
10. *A Thousand Tiny Cuts: Mobility and Security across the Bangladesh-India Borderlands*, by Sahana Ghosh
11. *Where Cloud Is Ground: Placing Data and Making Place in Iceland*, by Alix Johnson
12. *Go with God: Political Exhaustion and Evangelical Possibility in Suburban Brazil*, by Laurie Denyer Willis
13. *Kretek Capitalism: Making, Marketing, and Consuming Clove Cigarettes in Indonesia*, by Marina Welker
14. *Tabula Raza: Mapping Race and Human Diversity in American Genome Science*, by Duana Fullwiley
15. *Life at the Center: Haitians and Corporate Catholicism in Boston*, by Erica Caple James
16. *Zainab's Traffic: Moving Saints, Selves, and Others across Borders*, by Emrah Yıldız
17. *How to Love a Rat: Detecting Bombs in Postwar Cambodia*, by Darcie DeAngelo
18. *A Burdensome Experiment: Race, Labor, and Schools in New Orleans after Katrina*, by Christien Philmarc Tompkins
19. *Tackling the Everyday: Race and Nation in Big-Time College Football*, by Tracie Canada
20. *Kigali: A New City for the End of the World*, by Samuel Shearer

Kigali

A New City for the End of the World

Samuel Shearer

UNIVERSITY OF CALIFORNIA PRESS

Open access edition funded by the National Endowment
for the Humanities.

University of California Press
Oakland, California

© 2025 by Samuel Shearer

This work is licensed under a Creative Commons (CC BY-NC-ND) license.
To view a copy of the license, visit http://creativecommons.org/licenses.

All other rights reserved.

Suggested citation: Shearer, S. *Kigali: A New City for the End
of the World*. Oakland: University of California Press, 2025.
DOI: https://doi.org/10.1525/luminos.250

Library of Congress Cataloging-in-Publication Data

Names: Shearer, Samuel, author.
Title: Kigali : a new city for the end of the world / Samuel Shearer.
Other titles: Atelier (Oakland, Calif.) ; 20.
Description: Oakland, California : University of California Press, [2025] |
  Series: Atelier: ethnographic inquiry in the twenty-first century ; 20 |
  Includes bibliographical references and index.
Identifiers: LCCN 2025011314 (print) | LCCN 2025011315 (ebook) |
  ISBN 9780520423183 (cloth) | ISBN 9780520409972 (paperback) |
  ISBN 9780520409989 (ebook)
Subjects: LCSH: City planning—Rwanda—Kigali—History—21st century. |
  City planning—Climatic factors. | Sustainable urban development—
  Rwanda—Kigali—History—21st century.
Classification: LCC HT169.R952 S54 2025 (print) |
  LCC HT169.R952 (ebook) | DDC 307.1/2160967571—dc23/eng/20250616

LC record available at https://lccn.loc.gov/2025011314
LC ebook record available at https://lccn.loc.gov/2025011315

GPSR Authorized Representative: Easy Access System Europe,
Mustamäe tee 50, 10621 Tallinn, Estonia, gpsr.requests@easproject.com

34 33 32 31 30 29 28 27 26 25
10 9 8 7 6 5 4 3 2 1

For Mignonne and Shaya

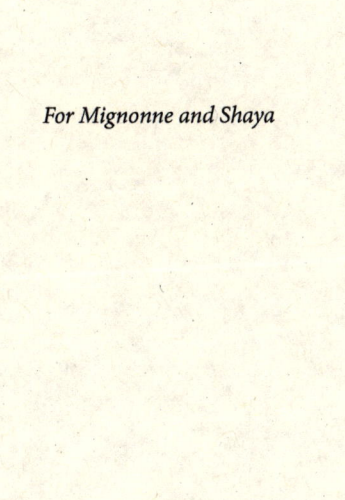

CONTENTS

List of Illustrations   viii
Preface   xi

1. Introduction: Capital of the African Century   1
2. Production: Making an African Metropolis   26
3. Brand: Ruled by Fictions   55
4. Destruction: Making Fictions Real   86
5. Repair: Punk Urbanism   113
6. Recycle: Wasted Space   142

Epilogue: Urban Humanities for a Broken World   169

Acknowledgments   181
Notes   185
References   193
General Index   211

LIST OF ILLUSTRATIONS

LIST OF MAPS

1. The sites that feature in this book   2
2. The growth of central Kigali   31
3. The OZ team's vision for the location of the New City Center   75

LIST OF FIGURES

1. Reconstruction of nineteenth-century Rwandan reed home with thatched roof at the King's Palace Museum in Nyanza   36
2. Postcard of Kigali's military camp during the Belgian occupation in 1916   36
3. Amatafari ya rukarakara drying before being used   38
4. A more recent title made through inzoga z'abagabo   43
5. Kiyovu cy'abakene before (*top*) and after (*bottom*) it was demolished in 2008   57
6. Gacuriro before it was demolished (*top*) and after the 2010 demolition (*bottom*)   58
7. Kimicanga before it was demolished (*top*) and after the 2012 demolition (*bottom*)   59
8. A rendition of the OZ vision of the New City Center   76
9. Amatafari ya rukarakara waiting to be used in a construction project   99
10. A single plot in Cyahafi, a neighborhood in the urban core   102

11. An image of the prototype Batsinda house as presented in the 2007 Kigali Conceptual Master Plan *103*
12. Where the Batsinda Housing Estate once stood, in 2023 *110*
13. A section of Dusheni that some call Dushenye ("[they] demolished us") *111*
14. Satellite images of Bannyahe in 2000 (*top*) and 2013 (*bottom*) *120*
15. The lower section of Bannyahe, from the path leading to it, in 2015 *125*
16. One way to cover up an X: with a discarded master plan printout *135*
17. The early morning caguwa wholesale in Biryogo in 2017 *143*
18. Abadozi workshop in Nyabugogo one year before it was demolished *144*
19. "The City Bazaar" *151*
20. The street taken over by Biryogo Market in 2017 *163*
21. The waste of sustainable urbanism: Nyabugogo Market in 2014 (*left*) and 2017 (*right*) *164*

Preface

... like every generation before, we live in the aftermath.
—STEVEN JACKSON (2014)

It was the right song at the right time:

Pas de retour! Pas de pardon! (No retreat! No surrender!)

Hatuwezi kurudi nyuma! (We never go back!)

Ni ukuri kw'imana mba mbaroga, nyabarongo! (I swear to God, I am telling you the truth, Nyabarongo!)

—AMAG THE BLACK, "NYABARONGO" (2014)

In October 2014 the Kinyarwanda hip-hop duo AmaG The Black and Safi Madiba uploaded to YouTube their hit music video "Nyabarongo." The song celebrated Kigali's popular cultures and spaces during some of the most controversial neighborhood demolitions in Kigali.

Within days, "Nyabarongo" was downloaded to thousands of flash drives by vendors in Kigali's street studios. The flash drives found their way onto the city's popular transport option, minibus taxis.[1] Taxi operators blasted the song over their passengers' voices as they traversed all three districts of Kigali. Bar owners used the song to attract patrons, and residents broadcast it from their homes. The chorus, "Pas de retour! Pas de pardon!," became famous.

In the video, AmaG expresses his trademark love for the streets. He employs his signature parody-as-critique, encoding the political message of the song in humorous inferences; code-switching between French, Kinyarwanda, and Kiswahili; and insider references to street culture—a strategy that keeps his music off the radar of the Rwandan government's tragically unhip (but scary) surveillance agencies. The latent message of "Nyabarongo" is this: You can't keep all the *abaturage* (regular folk) out of the city all the time. You can tear down our houses, ban our economies, and criminalize our homes and workplaces in zoning codes. But once we cross the Nyabarongo River—the unofficial border between the city of Kigali and the countryside—we never go back.

. . .

I am not from Kigali.

I came of age during the 1990s in a Colorado town that made lunch meat, methamphetamine, and one famous comedian. My most productive time during those years was spent skateboarding and running from the police—excellent training for a future career in the urban humanities. The best skate spots, however, were not in my hometown. They belonged to our much better-known neighbor: the mountain oasis college town of Boulder.

Many outside commentators on Boulder agree: there is something off about the city. Boulder is widely known as one of the most politically and environmentally progressive cities in the United States. Check out the City of Boulder's website—it is all about racial equity and inclusivity. And yet the city hosts very little actual racial or economic diversity. Eighty-nine percent of Boulder is white (US Census Bureau, n.d.), like one-percenter white (Williams 2008). And while it is difficult to locate the city's racism and classism in a single source, a subtle, but nevertheless pervasive, anti-Black (Cortina 2015), anti-unhoused (Becker 2022), anti-immigrant, and anti–working class (Brico 2017) sense lingers below the scent of patchouli oil, the sound of jam bands, and the bustle of vegan grocery stores.

In 1990s Boulder, skateboarding—especially street skating—was a crime. To be fair to city lawmakers, my friends and I broke stuff. A lot of stuff. Other people's stuff. We ruined marble ledges, sidewalk furniture, handrails, planters, you name it. If we wanted to skate it, we put Sex Wax on it,[2] grinded it, slid it, rock-and-rolled it, and blunted it until it crumbled. We broke stuff by accident, too, like the time I took out all the glass in a 7-Eleven storefront with a poorly landed manual.[3] But what made city officials in Boulder nervous about us was not so much what we did as who we were: foul-mouthed, punk-rocking truants from broken homes in Colorado's postindustrial hinterland. It was, after all, the nineties. There was a general moral panic over unsupervised, thrill-seeking adolescent anarchists with baggy pants and skateboards taking over the streets of US cities—just check out Larry Clark's 1995 handwringing docudrama *Kids*.

In the cartography of skate spots in Boulder, one stands out: the Pearl Street Mall. It is more street than mall—several blocks of car-free space in the middle of Boulder's downtown. It has everything you want to skate: sidewalk furniture, marble ledges, planters, rails, gaps, even faux cobblestone pavers that let out these wonderful clacks when you land a trick heavy. Pearl Street was (and still is) also a "bust"—like a 24/7, always-a-cop-onsite, zero-tolerance-for-anyone-who-doesn't-belong, capital *B* bust. So much for inclusivity.

When we were skating Pearl Street and we saw the cops, we ran. If one of us did get caught—and if the police could be bothered to play the "Now what's your *real* name?" game with us—we were always threatened with the same charge: trespassing. We never asked about the legal precedent for trespassing in public space. It was just how the division of labor worked. The police chased us. We ran. Everyone had their job to do.

Of course, the cat-and-mouse game my friends and I played with Boulder police was very different from the experiences of people who had their lives and neighborhoods turned upside down by stop-and-frisk and other late twentieth-century revanchist policing strategies that targeted minoritized communities in US cities. Despite being liberal democracies on paper, twentieth-century incentivized urbanism converted many cities across the United States into anti-Black, anti-immigrant, anti-unhoused cities run by zealous policing strategies. Still, these early experiences provided me with a partial view into the force that is necessary to defend the "best practices" of privatized, green city-building from its alternatives.

I eventually left my hometown for the bright lights of bigger cities and forgot all about Boulder and Pearl Street. Until 2007. That year, as a twenty-eight-year-old, late-to-bloom, slow-to-learn undergraduate at City College of New York, I ended up in Kigali, Rwanda, on a college work-study travel program. By happenstance, my first trip to Kigali overlapped with a day of surprise violence—the day when a section of the downtown Kigali neighborhood Kiyovu cy'abakene was demolished. No one but the city's managers knew it then, but that day was a dry run for an operation the following year that would wipe out the entire neighborhood, and the one next to it, and the neighborhood next to that—all in the name of a project called "sustainable urbanism."

After the demolition, Ephraim, a close friend of mine who has lived in Kigali all his life, took me to see the wreckage.[4] It looked to me as though a bomb had ripped through a section of the neighborhood. I saw people combing through the detritus of what used to be their homes, salvaging steel doors and household items.

"What happened?" I asked Ephraim.

He asked a few people nearby. "They say there's a master plan." He shrugged.

Later that year, back in New York City, I searched the internet for "master plan Kigali" and there it was: the Kigali Conceptual Master Plan, written by the Colorado firm OZ Architecture. I expected to see technical documents and maybe a narrative that would shed light on Kigali, its history, and the future direction of the Rwandan capital. I hoped for an explanation of the rubble I had seen in Kiyovu that day. Imagine my surprise when, instead of Kigali, the plan contained images of the Pearl Street Mall, the University of Colorado underpass, the civic center—all places my friends and I knew from Boulder's zero-tolerance busts.

The more I learned about the plan, the more I learned about Kigali, the more a suspicion from my adolescence was confirmed. The police who had chased us around Boulder were hired muscle for a project that was much larger than protecting concrete ledges from skater punks and extended far beyond our cat-and-mouse game.

The lead architect and planner on the Kigali Conceptual Master Plan was Carl Worthington, who was also one of the designers behind the Pearl Street Mall. Just after the Kigali Conceptual Master Plan was released, Worthington described Boulder as "my little laboratory of experimenting with urban-design things all these years" (*5280 Magazine* 2008). In his laboratory, Worthington did more than design. He sat on Boulder's municipal policy boards and was instrumental in creating Boulder's famous greenbelt. It was also during this time when Boulder's municipality passed its 1976 comprehensive master plan with zoning codes that used the language of nature conservation to justify other projects. "Dirty" industries like meat processing (and the people who depended on these industries) were zoned out of the city. "Clean" industries like technology, academic research, and finance were promoted. As a result, property values soared to some of the highest in the nation, and Boulder became a green lifestyle city.

As Boulder's official policies became more open, more progressive, and less overtly racist, the technical language of nature conservation was used to defend Boulder's history as a white settler enclave of urban privilege, continuing a project of anti-poor exclusionary city management by policing its alternatives (Hickcox 2007).

For my friends and me, street skating was just one of a series of related cultural practices that were about trying to do something with the ruins of postindustrial US cities: hip-hop block parties, anarchist squatting, street art. It was our best, if imperfect, attempt to make life a little less alienating in the aftermath of the urban catastrophe that is now called neoliberalism. In Kigali, neighborhoods like Kiyovu cy'abakene and the architectures and economies that supported it had emerged as collective responses to the catastrophes of cold-war capitalism, civil war, and genocide. As alternatives to privatized green city-building, they posed a threat to the particular vision of Kigali's green future that the government of Rwanda had purchased from outside firms like OZ Architecture.

...

Over the next decade and a half, I studied Kinyarwanda, finally got accepted to graduate school, and spent more time hanging out in Kigali streets with insurgent architects, infrastructure builders, and street traders—people who also spend a lot of time running from the police. I followed the technologies of privatized, green city-building as they traveled from the United States and Singapore to Kigali. I watched neighborhoods like Kiyovu cy'abakene, Gacuriro, Kimicanga, Kosovo, Lafreshaîre, Bannyahe, and many more destroyed in the name of sustainable urbanism. I learned about the unconventional ways Kigali residents repaired and rebuilt the popular city from the detritus of green capitalism while they pushed back against corporate urban planning with designed worlds of their own.

It bears repeating: In recounting my path to this book, I am in no way suggesting that my teenage adventures on a skateboard are comparable to the violence of dispossession that many Kigali residents have experienced as a result of market-driven sustainable urbanism. As a white US academic with a PhD, I am structurally closer to the consultants who wrote the Kigali City Master Plan. But the pages that follow are not about me. They are about how Kigali residents repair urban worlds in the aftermath of the disasters of colonialism, genocide, and green capitalism. They offer a case for questioning the seemingly unquestionable project of finance-driven sustainable urbanism as the only urban response to climate change. And they suggest that we (humans interested in the survival of humanity on a broken planet) have much to learn from the often-silenced popular alternatives to finance-driven sustainable urbanism—the imperfect processes of collaborative survival and social engagement that are found in Kigali.

So there you have it: why I—a white guy from Colorado who is now a middle-aged, middle-class academic—chose to write this book, how I came into it, and why it is written from Kigali streets. Finally, I am not an expert on all things Kigali or Rwanda, as if such a person could exist. I understand my role as akin to what the cyberpunk writer William Gibson (2012, 93) calls being "a sort of butler." I want to open doors for you, invite you in, and share with you what I have learned about sustainable urbanism and the production, branding, destruction, repair, and recycling of urban space from the streets of one of the most significant—if misunderstood—cities in the world.

Welcome to Kigali: capital of the African century.

1

Introduction

Capital of the African Century

DECEMBER 2011

The mayor of Gasabo, the largest district in Kigali, Rwanda, orders thousands of landowners and tenants to leave Kimicanga, a third-generation, working-class, mixed-use, pedestrian-only neighborhood within walking distance of the city's central business district. Over the next year, thousands of homes, dozens of bars, several cultural centers, and a large food and clothing market are destroyed by the city's bulldozers. Evictees from Kimicanga end up scattered throughout the city, mostly on Kigali's edges. About fifty households, however, decide to stay near the city center. They take their chances near the wetlands, where they purchase land from farmers and hastily build a neighborhood they call Kosovo—a cosmopolitan nod to a famous "slum" in Nairobi that is named after the once war-torn southern European nation.

While Kimicanga is being torn down and Kosovo is coming up, another project is being installed across the wetlands: KN 123—twenty kilometers of steep tarmac road that begins in a working-class neighborhood at the top of Nyarugenge Hill and winds down through the exclusive, upscale neighborhood of Kiyovu. The road terminates at a roundabout at the bottom of the hill, just across the wetlands from Kosovo. To protect the road from erosion, construction workers push hundreds of perforated PVC pipes through the stone and into the backfill that supports the road. These pipes drain stormwater from the hillside and drop it into cement gutters that run downhill along the length of the road; ultimately, the water drains into the wetlands below Kosovo. The road takes two years to construct; it is completed in December 2013.

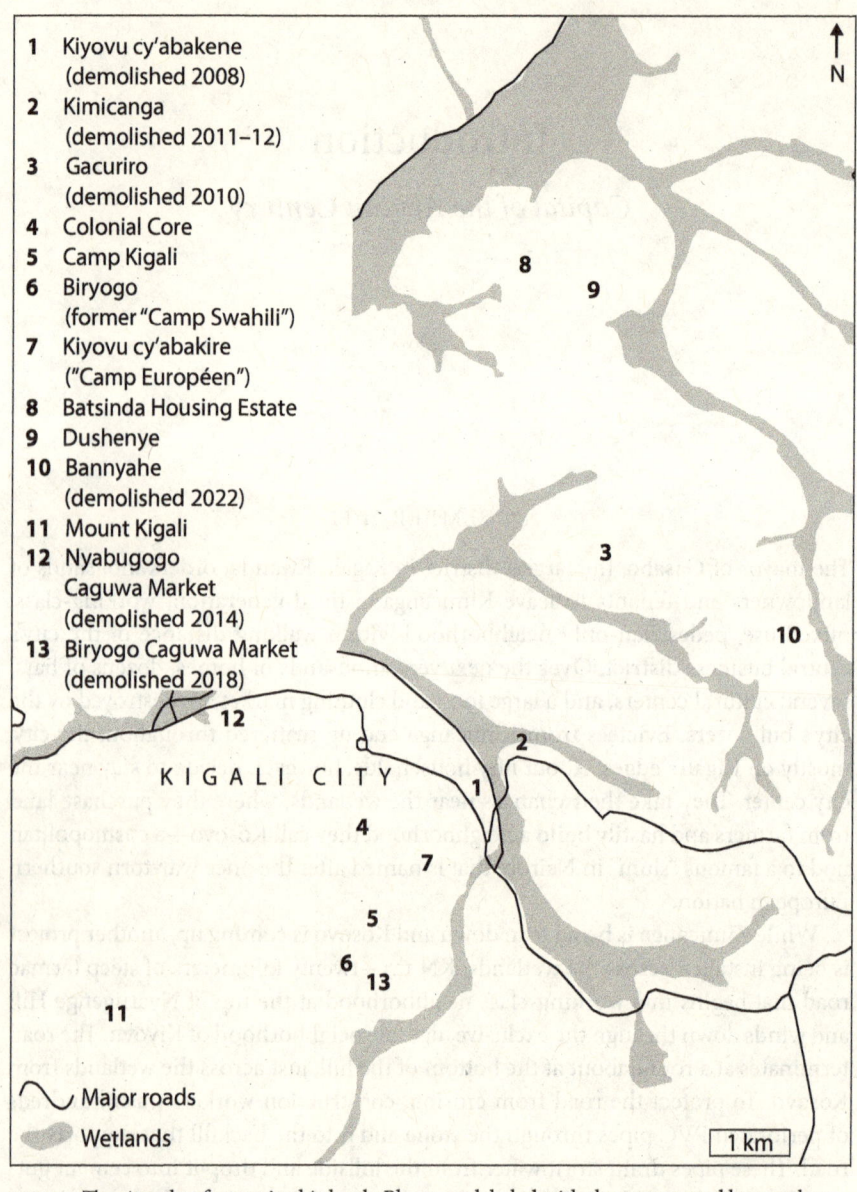

MAP 1. The sites that feature in this book. Places are labeled with the names used by people who live there (which are sometimes monikers) instead of the official names used by no one but city authorities.

DECEMBER 17, 2013

When the rain begins, the city's drainage infrastructure holds its own. Two hours into the storm, however, trouble begins at the roundabout where KN 123 ends—where the water that is drained from the road's backfill flows into the wetlands. The clay pipe that, for several decades, has directed stormwater safely below the roundabout is unable to handle the additional drainage from KN 123.

The pipe overflows, and stormwater floods the roundabout. The road collapses under the pressure, shattering the drain below. Pieces of tarmac and clay are carried away in the resulting flash flood. Freed from the city's drainage network, water rushes downhill toward the wetlands. As the water flows into the wetlands, water levels there begin to rise—up toward Kosovo, where dozens of families evicted from Kimicanga have settled.

"Jesus Christ! It was like the apocalypse!" (*yesu we! wagira ngo ni imperuka!*), remembered Ms. Iradukunda, one of the evictees living in Kosovo that day.[1] "Because all that water in just one day," she said, shaking her head, "... the water took everything—the house, all my possessions were swept away. We had to take refuge in Restoration [a church uphill] for the next week" (recorded interview, February 3, 2015).[2]

In the end, the water destroys more than forty houses. Their contents are buried, in the words of Ms. Iradukunda's neighbor and former landlord, Mr. Nsengimana, "beneath the mud and stone" of the flood sediment.

• • •

This book is an ethnography of a city that is being destroyed so that it can be rebuilt for the end of the world. It is about popular cultures and street economies where there is no public space; the production and defense of popular urbanism where the government has zero tolerance for "disorder"; and the ruination of built environments in the service of a project called "sustainable urbanism." I follow these contradictions as they unfold during a twenty-first-century experiment with utopia, green capitalism, and extrastatecraft: the Kigali City Master Plan.[3]

Although it is often spoken about by city officials, designers, and media boosters with a definite article, "the plan" is not a singular project. It is, rather, a series of multiscalar, international ventures in financing, database creation, marketing images, and government initiatives aimed at converting Kigali from a postconflict "wounded" city into an optimized hub of global finance and tourism that will withstand the shocks of climate change. Owned and implemented by the government of Rwanda; written, revised, and rewritten again by design firms in the United States and Singapore; and financed by private capital and public debt, the Kigali City Master Plan is also inspired by the UN agenda to reimagine the city as a capitalist solution to the dual crisis of economic and ecological catastrophe.

Focusing on the related projects of branding, destruction, repair, and recycling, this book shows that demolition, dispossession, and infrastructurally induced catastrophes, like the demolition of Kimicanga and subsequent floods in Kosovo, are not the unintended consequences of technocrats trying to solve the problems of ecology and economy and then failing. Rather, I argue that sustainable urbanism, as it is practiced in Kigali, is the destruction of built environments and the conversion of crises (real and imagined) into new markets for expert labor, green commodities, and real estate. That is why this book begins with the end: a planned apocalypse in which human activity combined with an abnormal weather event to manufacture a housing shortage and push residents into the wetlands, ending in a "natural" disaster that left much of the urban core in ruins. These processes make real marketing fictions about Kigali as a postcrisis city; about African cities *in crisis*; and about green capitalism as the utopian solution to the end of the world.

Much of what I claim to know about sustainable urbanism, the city, and green capitalism in Kigali is built on ethnographic work carried out over ten years, mostly done during an uninterrupted period from May 2013 to August 2015 and buttressed by two summers of prefieldwork visits and return trips to the city every year until 2023. This research was done in Kinyarwanda, a language I speak, read, and write. I also draw from a rich archive of contemporary Kinyarwanda popular expressive culture (music and film), Kinyarwanda print media and online journalism, and Africanfuturist renderings of urban life, as well as satellite images of the city.

The people who give form and content to the chapters that follow are fashionistas, construction workers, middle-class landowners, tenants, secondhand clothing traders, small shop owners, street traders, and construction workers. They identify as women and men, young and old. They represent three generations of mostly Rwandans, but also Congolese, Tanzanians, Burundians, and Ugandans who have made the Rwandan capital their home. This cosmopolitan collective of city producers—and the spaces they make and live in—does not exist in the documents and marketing materials that make up the Kigali City Master Plan. They are the alternatives to a project that claims to have no alternative, Rwandans who rebuilt their own city after the 1994 genocide against the Tutsi.[4] They plan and build accessible and affordable urban spaces with minimal impact on the city's ecologies, despite claims by US academics that Africans "lack the historical agency" to do so (Davis 2004, 2006). They are, in the words of the late Black studies scholar and anthropologist Michel-Rolph Trouillot (1995), "unthinkable" within the categories of "slum," "informality," and "development" that structure discourses on African cities. And they are the subjects and spaces that are erased and rendered "ungeographic" (McKittrick 2006), and therefore insignificant, in the process of theory production.

Ms. Iradukunda, for example, is not a recognized expert on sustainable urbanism. She is a single mother of one and a jill-of-all-trades. She sells produce in discount street economies, works overnight construction on buildings without permits, and picks up night shifts in unlicensed bars, all activities that land her from time to time in Kigali's infamous extralegal detention centers under vague charges of *akajagari* (unauthorized "disorderly" commerce). But her knowledge of sustainable urbanism, coming as it does from a lived, embodied experience, rather than statistics and planning models, makes her an excellent interlocutor.

After Ms. Iradukunda recalled her memory of the flood as *imperuka* (the apocalypse), she invited me to step outside to see where it all happened. We did not need to go far, only a hundred yards downhill on a sandbagged footpath that she and her neighbors had stamped into the clay earth during *umuganda* (monthly compulsory community service days). She had been moved three times in five years because of master plan projects. Each displacement was aimed at relocating her and her neighbors to homes that were too small, located in *imidugudu* (villages) on the city's edge where no one wanted to live. Refusing to go to the imidugudu, she decided to stay, a few hundred yards from where her life in the city had begun.

"And here," Ms. Iradukunda said, "they say they are going to demolish this place, and we will go." She laughed.

"Where? To the imidugudu?" I asked.

More laughter. Ms. Iradukunda raised her eyebrows and looked at me in mock surprise. Anywhere but the imidugudu.

Then, she asked, "Isn't Kigali big [*Kigali si hanini se*]?" She answered her rhetorical question herself: "It's big. If you want a place, you go there" (recorded interview, February 3, 2015).

If you want a place, you go there.

As Kigali's teams of managers set about destroying the city so it can be rebuilt as a privatized green metropolis, Kigali residents push on the world of sustainable urbanism with designed worlds of their own. Popular Rwandan architects are going *kugikoboyi* (cowboy), rebuilding low-impact, low-carbon, ordinary neighborhoods from renewable materials and the discards of demolition with names like Kosovo and Banya/bitmahe (where do they piss/shit?) and Dobandi (for gangsters), against Kigali's brand image as the "cleanest, greenest city in Africa." Itinerant "informal" street traders address their criminalization by producing nomadic popular spaces and discount economies that migrate around the city, staying put by moving around. And Kigali residents recycle the detritus of demolished built environments and discards from other places into popular fashions, media, and spaces that frustrate city managers' attempts to "get the city in line" (Simone 2009, 7). These are not dichotomous processes but the outcomes of efforts to convert the value of popular economies and technologies into new markets for green capital while matching the city to representations of crisis.

A key issue, for example, in studying the relationship between urban planning and daily life in Kigali is the yawning gap between the contents of the master plan and the activities urban managers carry out in its name. A neighborhood like Kimicanga is demolished, and everyone—city authorities, the evictees, the Kinyarwanda press—agrees that it happened because of the plan. But none of the plans to build a sustainable future in Kigali include a program to destroy Kimicanga, or any neighborhood. Rather, the consultants who write plans for Kigali express a commitment to economic growth through equity, or the distribution of resources across economic classes. They warn *against* neighborhood demolition as *unsustainable* because it disrupts communities and social networks. Yet between 2007 and 2022—the period covered in this book—thousands of homes and workplaces and dozens of communities in Kigali were destroyed, and thousands of people were displaced, all in the name of sustainable urbanism.

A second, related problem is the lack of available categories and perspectives in orthodox policy studies to describe neighborhoods like Kimicanga. Writing about the Kimicanga demolitions ex post facto, other scholars reproduce official accounts. Kimicanga must have been demolished because it was an "informal," or "slum," settlement, concepts that indicate a general sense of lack—lack of form, lack of adequate housing, and lack of security.[5]

But anyone who spent time in Kimicanga before it was demolished or spoke with people who used to live there, like Ms. Iradukunda, knows that the qualities we assume to be absent in "slums" were present there.[6] The neighborhood was centrally located and sunk into the city's infrastructural base. For three generations, documented property owners controlled a thriving affordable housing market that supplied thousands of units to the city's stock of dwellings, mostly built with materials—clay, earth, sand—from the wetlands below. It hosted popular economies based on discounting and moving produce that was about to go bad. Hospitals and schools were within walking distance, as were jobs, and residents who lived there had access to popular transportation.

In addition, as an entrée to Kigali from elsewhere in the region, Kimicanga had become a cultural center that brought musicians and artists from Congo, Burundi, and elsewhere in Rwanda into the same space. It was a home base of *igisope*, a fusion of funk, Kinyarwanda folk, and reggae that rocked famous neighborhood clubs like Cadillac. People congregated and did business at sports bars like Kwa Monica. Kimicanga was not perfect. It had its share of health and environmental issues. Its proximity to the wetlands made the lower part of the neighborhood flood prone, and it had its share of issues with inequality and intermittent services like garbage pickup. In the words of people who used to live, hang out, and dance in Kimicanga (myself included), the area was *haciriretse* ("ordinary"; also "normal," "average"). And that was precisely the problem.

Standing, Kimicanga and the dozens of other neighborhoods demolished in Kigali since 2008 posed a challenge not to the "world-class" aesthetics of the future luxury city but to the representation of Kigali in the master plan as a series of profitable problems to solve. As an ordinary neighborhood, built by Kigali residents after colonialism and then repaired and extended after the 1994 genocide against the Tutsi, Kimicanga posed a threat to the marketing narratives that frame Kigali as a postcrisis city of "informal settlements" waiting to be rebuilt by external investors. The neighborhood provided evidence that the Kigali City Master Plan's numbers, imported from UN-Habitat, that count 83 percent of the city as living in "informal settlements" and therefore a market for tiny houses, rainwater catchers, and biogas converters were incorrect. It was evidence that Kigali residents—Africans—were capable of envisioning, planning, and building accessible, low-carbon urban spaces, popular institutions, economies, and infrastructures without the assistance of foreign investment or experts from the United States and Singapore. In other words, Kimicanga was not demolished because it was a "slum" and a local elite found it to be a "nuisance" (see Ghertner 2015). Kimicanga was demolished because it was *not* a "slum" and was therefore unthinkable within the narratives of crises in the Kigali City Master Plan.

"When reality does not coincide with deeply held beliefs," Michel-Rolph Trouillot writes, "human beings tend to phrase interpretations that force reality within the scope of those beliefs" (1995, 72).

The ethnographic project of this book is to make the lives, spaces, technologies, and economies that are unthinkable within the confines of top-down planning in Kigali thinkable, as viable, African alternatives to green capitalism. My project is inspired by other scholars in the African humanities who argue that the continent should be seen "not only as a site of ecological degradation" but also as a source of "inspiring epistemological models," artistic expressions, and ecological thought (Iheka 2021, 10; see also Mutongi 2017; Myers 2016). This book therefore is neither a story of miraculous recovery nor a tale of melancholic suffering from a place that, for the last thirty years, seems to have inspired only such accounts. It is a book about life in a city that is being broken, repaired, and recycled—in ways that are sometimes spectacular and in other ways mundane—by the "best practices" of twenty-first-century sustainable urbanism. My underlying political project is to critique green capitalism as more than a bad idea—as a destructive process that accelerates inequalities and the effects of climate change. Along the way, I aim to demonstrate the value of imperfect spaces and practices that Kigali residents have made themselves to demonstrate who pays the costs for producing Kigali's green futures.

CAPITAL OF THE AFRICAN CENTURY

There are many reasons for writing this book from twenty-first-century Kigali. The Rwandan capital is one of the fastest-growing cities in the world and is often cited by the United Nations, World Bank, and international award-granting architecture associations as a "model" metropolis for Africa (Bafana 2016). Consequently, foreign consultants who work for global design firms see Kigali as an ideal place to experiment with incentivized city-building projects. The city is run by a ruling party that is also a global investment firm—what its Wall Street boosters in the United States call, without irony, "Rwanda Inc." (see Crisafulli and Redmond 2012). Unlike neighboring cities where democracy is also in question, the ruling party of Rwanda has a reputation for getting things done with frictionless, streamlined pro-business precision. And while Kigali is far from the only city in the region with a new master plan, the city is nevertheless influential in policy and model building throughout the continent. Rwanda, the South African *Mail & Guardian* reminds its readers, preempting the thesis of the epilogue to this book in a very different way, "is your future, whether you like it or not" (Allison 2017).

Of course, Kigali, a capital city of somewhere between 1.7 and 2 million people, depending on whom you ask, is not as unique as the marketing hype claims it is. Despite sophisticated public relations strategies, Kigali is also an "ordinary" city in many senses of the term (Robinson 2006). Like many other cities in the Global South, Kigali is host to large numbers of residents who live and work outside formal wage employment, and most of the city is built outside of formal regulations despite heavy-handed efforts to impose zoning and building codes. Further, Kigali must compete for "capital of the African century" status with other cities in the region that have their own plans for privatized green futures. In Dar es Salaam, Addis Ababa, Accra, Dakar, Luanda, Kampala, and elsewhere, local governments are razing neighborhoods and offering land and tax incentives to investors to build green enclaves, airports, seaports, innovation districts, and upscale housing estates (see Myers 2016; Gastrow 2024).

What makes Kigali appear unique, at least in booster accounts of the master plan—how the city's brand managers present it as an irreplicable opportunity to build a green technofuture—is that Kigali already occupies a place in the global imagination that scholars of the postapocalypse call "after the end" (Berger 1999): a postconflict, postgenocide, postcrisis city, lying ready for planners and external investors to build a new city designed to withstand the effects of climate crises.

This book addresses the memory of the 1994 genocide against the Tutsi without reducing the built environment to that event alone. I attend to the ways foreign consultants, Rwandan authorities, and international finance institutions leverage the image of Kigali as a postcatastrophe city, rewriting the traumatic event as a unique opportunity to build a future-proofed metropolis. But this book is also intended as a challenge to academic work, journalism, and popular culture, mostly produced in the United States and Europe for Western audiences, that reduces everything that happens in Rwanda to a case study of mass violence and postconflict recovery. Indeed, part of what makes this book original is that it refuses to treat Rwanda as a place to extract melancholic memories for sale on the Western trauma market. Instead, I write of Kigali as a contribution to a growing scholarship in the urban humanities from Africa.

Ms. Iradukunda's narration of the December 2013 flood and the events that led up to it outlines the theoretical interventions of this book: the contradictions of finance-driven sustainable urbanism and the destruction it brings; the power of dys/utopian city marketing narratives; and the ways Kigali residents repair their own city in the wake of catastrophes. Her choice of *imperuka* (apocalypse) to describe the flooding evokes a subgenre of dys/utopian storytelling with specific plot points—the end, after the end, and a new beginning. And it justifies the interdisciplinary methods I harness in this book: an urban humanities that brings together ethnography and Africanfuturist perspectives on the city in a study of sustainable urbanism and its alternatives.

In what remains of this introduction, I outline the methodological and theoretical contributions of the book and map their place in the chapters that follow. After briefly outlining the origins of sustainable urbanism in Kigali and defining what I mean by green capitalism, I turn to a general discussion of city branding, dys/utopia as narrative genre, and the politics of repair. This leads to a discussion of the urban humanities and Africanfuturism as method, followed by an overview of the book's structure.

THE ORIGINS OF SUSTAINABLE URBANISM IN KIGALI

The project of sustainable urbanism, as it is practiced in Kigali, is not to get the city "right." It is to sell Kigali to hedge funds, tourists, parastatals, real estate firms, and other investors.

Selling Kigali to the world became official government policy in 2002. That year, the mayor of the city of Kigali signed the Kigali Economic Development Strategy (KEDS), a collaboration between the government and researchers at the Kigali Institute of Science and Technology Studies (KIST). The purpose of KEDS was to address issues that were simultaneously particular to Rwanda's recent history and representative of a general "trend in developing countries" everywhere (City of Kigali 2002, 7): Kigali's population was booming, but thanks to a crashed postgenocide economy, the city's revenue was not.

The KEDS report describes the problems in Kigali at the turn of the twenty-first century: roads needed to be fixed; education, health, and utility services needed to be extended. To solve these problems without a tax base, the report's authors proposed bringing money into the city through foreign direct investment. To achieve this goal, the city needed to "aggressively" develop "a professional campaign to market Kigali to foreign investors" (City of Kigali 2002, 87). To attract those investors, the authors of KEDS wrote, the city needed to change its image (see City of Kigali 2002, 70).

A year later, in summer 2003, the government of Rwanda and investment-driven sustainable urbanism met each other in Aspen, Colorado, at a Fortune 500 event, where Rwandan president Paul Kagame was introduced to architect and planner Carl Worthington, of the Colorado-based architecture firm OZ.

President Kagame had been in office for only three years.[7] Worthington would later describe the meeting as a small firm with big ideas meeting a president of a small country with a big vision. "I have worked in Cape Town at a resort complex," he said, "but this is my first in terms of a major urban development in central Africa. It is a tremendous opportunity, to create a new model of a democratic, ecological, high-tech city that would become a symbol of the new Africa" (K. Mitchell 2005, 1).

According to the interview, shortly after he met President Kagame in Aspen, Worthington demonstrated how his firm operated, giving the president and his cabinet ministers a tour of one of his signature projects, the Denver Tech Center, a private technology park built in the late 1960s on vacant land south of Denver, Colorado. After the tour, Worthington told his interviewer, "I presented a slide show of what the DTC looked like in 1962—rolling grasslands with a 40-acre vision that grew into an important, mixed-use, high-tech ecological center" (K. Mitchell 2005, 1). President Kagame and his cabinet, Worthington said, "began to see that we also had visions and were able to carry them out without big government funding. You find the money" (K. Mitchell 2005, 1). Kagame would later say in his own interviews that rebuilding Rwanda through foreign investment was part of an economic strategy of national self-determination. The idea, later popularized by the Zambian economist Dambisa Moyo in her book *Dead Aid*, was to replace foreign aid with foreign investment. Finance would, in theory, delink Rwanda from the donor industries and the political strings attached to aid. In the process, finance would, at least in theory, free Rwanda from its dependent relationship on Western development industries by replacing aid with foreign investors who cared only about profit (Moyo 2009, 27–28).

The ruling party of Rwanda wanted to develop a new image of Kigali to attract foreign investment. To achieve that goal, they needed a firm like OZ: intermediate actors who could be compelling to outside target audiences. The OZ team correctly identified that those investors wanted a quality called "sustainability," which, the OZ team writes in its first plan, "can only enhance Rwanda's attractiveness to foreign direct investment and other forms of economic development" (OZ Architecture 2007, 12). While the concept of sustainability had been around for some time, the OZ plan responded to a shift in the power of sustainability that took place in the first decade of the twenty-first century: the idea that capital could be reengineered to save both the environment and capitalism.

SUSTAINABLE URBANISM AS GREEN CAPITALISM

Kigali went sustainable during the first decade of the twenty-first century. It was "the decade when global climate change became a salient international issue" (Death 2015, 2210; see also Checker 2020, 36). The Kyoto Protocol and a series of globally visible climate change–related urban disasters coincided with the 2007–8 global financial meltdown, leading to calls from global governance and corporations to reallocate finance capital away from fossil fuels and toxic derivatives and toward renewable energy and green infrastructures. Slogans like "sustainability" and "the green economy" went from being buzzwords used by the United Nations to concepts "deployed to legitimize and accelerate market-oriented growth and the social and ecological contradictions that accompany it" (Greenberg 2015, 127). The word *sustainability*, once used as a critique of the effects of unlimited growth on the environment, had come to mean the opposite: a synergy of environmental rhetoric and unlimited returns on investments.

The concept of green capitalism names the contradictions that arise in efforts to clean up the environment with technical fixes while keeping all the inequitable features of free-market capitalism intact (N. J. Fox 2022; see also Holleman 2018, 75). More than greenwash, green capitalism refers to serious efforts by architects, entrepreneurs, and engineers to green (as a verb) polluting cities by using new tech to clean up construction, energy, waste streams, and, by proxy, capitalism itself (see Checker 2011; Goldstein 2018; Günel 2019; Myers 2016). Much critical scholarship on twenty-first-century sustainable urbanism shows how the greening of cities delinks the politics of environmental and social justice from climate change through what Gökçe Günel, in her work on Masdar City, calls "technical adjustments," as though "climate change is a management problem that experts may resolve rather than an ethical and moral problem that humans around the world should recognize, discuss, and address as political agents" (2019, 10; see also Swyngedouw 2009). Others call attention to contradictions of seeking growth-based solutions to ecological crises that were created by capitalism—what Melissa Checker, in her work on sustainable urbanism in New York City, calls "the inherently contradictory promise... that we can stimulate economic growth while mitigating the effects of climate change without any sacrifice" (2020, 7). These efforts to generate a synergy between market growth and the environment are more than just bad ideas waiting to be deconstructed. As Miriam Greenberg warns in her work on New York City and New Orleans, "insofar as a market-oriented discourse of sustainability becomes a dominant and powerful agent within contemporary capitalism and capitalist urbanization, it has the capacity to render other, non-market goals—whether ecological or social—unsustainable" (2015, 107–8).

Most work on finance-driven sustainable urbanism comes from cities that—unlike Kigali—have a massive carbon footprint that warrants intervention. This work focuses on efforts to clean up "dirty" industries in ways that accelerate market growth in efforts to save capital from the crises of its own making. *Kigali* shows that the process also works the other way around. In Kigali, as in many (not all) African cities, proponents of green growth argue that because the city has no history of industrial capitalism and because it has a tiny carbon footprint, it is the ideal place to install green service delivery systems without the messy business of cleaning up what is already there (see Carmondy et al. 2023; Hudani 2020). As I show in the chapters that follow, most sustainable projects in Kigali are not about saving the environment with clean and green tech. They are about converting already existing low-carbon materials and technologies—air-dried bricks made with local clay, low-waste fashion industries, and collective saving associations—into capital through the alchemy of expertise.

Converting already existing city-building practices into capital while converting residents into consumers for green commodities requires an enormous amount of work. Entire neighborhoods like Kimicanga that—despite having their share of issues—were once walkable, built with low-carbon materials, and run on discount economies of recycling must be demolished to generate the scarcity required to sell new affordable houses, solar panels, and rainwater catchers (see chapters 4 and 5). Market space is destroyed in efforts to convert recycling industries into interest-bearing capital for master plan projects (see chapter 6). In the process, "sustainable" interventions in Kigali's built environment, like the demolition of Kimicanga that put Ms. Iradukunda in the path of the flood described above, are making the city more—not less—risk prone, accelerating the unequally distributed effects of climate change on the city and the people who live there. City managers often frame these issues not as failures but as the conditions of possibility for green capitalism and the planning-as-marketing processes that support it.

BRANDING THE CITY

Kigali is not the only city in Africa with a new plan for a sustainable future. In the second decade of the twenty-first century, journalists and academics began to note the appearance of glossy publications and YouTube videos advertising future sustainable-city makeovers and enclaved, green "cities-within-the-city" for places as different as Kinshasa, Lagos, Kigali, Nairobi, Cairo, Johannesburg, and Accra. Some scholars engaged ironically with the visual aesthetics of these designs (e.g., de Boeck 2011). Others criticized these "urban fantasies" as inappropriate for their host cities (e.g., Watson 2014), and still others described these plans as potential "smart solutions" to Africa's "problems" of housing and infrastructure (CNN 2013).

Few, however, paid attention to the effect the continent-wide shift toward privatized sustainable urban futures built with foreign direct investment had on the relationships between cities across the continent. Once a municipality commits to nurturing foreign investment rather than developing a tax base, it commits to a specific theory of the city. The city becomes a commodity to be sold alongside other cities in a regional market (Easterling 2014; Goldman 2011; Harvey 1989; Rutheiser 1996). As consultants from the United States and Singapore remind their stakeholders in Kigali, they are not just imagineering new technofutures.[8] They are engaging in high-stakes competition with other cities in the region—all of which have their own master plans promising investors high returns on nearly identical green futures.[9]

The response to these competitive pressures proposed by the consultants who first worked on the Kigali City Master Plan was to develop a "brand" (their word) that would make Kigali stand out above its competitors.[10] Kigali's managers needed to identify what their audience of external investors saw as the city's irreplicable qualities—its "monopoly rents" that could support an authentic brand and would not be subject to the competitive pressures of the marketplace (see Harvey 1989). Those qualities provide, Miriam Greenberg writes in her influential work on city branding, the foundation of a brand image that is carefully constructed so that "the name of the city alone . . . conjure[s] up a series of images and feelings, and with them an impression of value" (Greenberg 2008, 34).

In what is often called "entrepreneurial" (competitive) city building, branding serves several functions. First, under the competitive pressures of investment-driven urbanism, branding provides the themes for advertising, which takes the form of master plans. These themes take the form of images, sometimes appearing as blueprints and maps, and reference a spectacular future; they allow the city's name to travel with its authentically unique qualities attached, establishing connections with investors, tourists, and media boosters.

In addition, urban brands, like commodity brands, serve as independent financial instruments that hedge against the risks of market fluctuations and falling profits, shifting competition "to the unique value addition of brand identity" (Nakassis 2013, 118). This hedging is particularly important when cities face regional competitive pressures for investors. An economic trade zone, tax incentives, and free land are only competitive until the neighboring city also offers these perks. But the unique value of the city's name and reputation, if managed correctly, can remain stable even in the face of competition in other areas.[11]

The separation of the value of the commodity from the value of its brand is reflected in the division of labor around the marketing of commodities: the highly paid design work of branding and marketing is done in financial centers in Europe, the United States, and Asia, and the low-paid manual labor of producing the commodity—and, in the case of the commodified city, paying the price for foreign direct investment—is done by unseen laborers in the Global South. The process of branding Kigali was no different. In fact, those who did the branding work had little connection at all to their product. Carl Worthington later remarked that, when he was first hired by the Rwandan government to work on Kigali's image issues, he "couldn't quite remember where Rwanda was" (City of Boulder 2013).

OZ, and the other firms that followed, did not need to know where Rwanda was. They needed to identify the qualities that would make Kigali appear unique in the eyes of their target audience of foreign investors and convert those qualities into an image that could direct utopian energy toward the Rwandan capital and away from the cities it was competing with. What is one attribute that travels with the name Kigali, presenting itself as a unique, irreplicable essence?

Whether fact or fiction, most cultural products about Rwanda made for foreign audiences leverage an image of genocide, not as a crime against humanity but as a general condition of Rwanda.

UTOPIA AND THE POSTAPOCALYPSE

From April to July 1994, the world watched as between eight hundred thousand and one million Rwandans were murdered by their neighbors, officials, friends, and, in some cases, relatives. The 1994 genocide against the Tutsi affected the lives of Rwandans everywhere and deserves the attention it has received.

The world abandoned Rwanda in 1994, but it also really saw Kigali for the first time. Much of the violence in the city was carried out in the open, in view of foreign correspondents who broadcast real-time images of mass violence to audiences back home. People all over the world who had never heard of Rwanda, let alone Kigali, were confronted with the unspeakable in their living rooms. Then came the aftermath—images of millions of refugees fleeing both sides of the war.

Three decades later, thanks to culture industries, Western journalists, and academics who seem unable to write about anything else when it comes to Rwanda, non-Rwandan audiences can be counted on to hear *Rwanda* and think one word: *genocide*. Regardless of what a news report, film, or documentary is ostensibly about—the Rwandan national cycling team, fashion, urban planning—if it is related to Rwanda and targets an outside audience, the framing device must be genocide and postgenocide rebuilding. This makes Kigali a convenient setting for Hollywood blockbusters, HBO docudramas, and major news outlets to explore the big issues of redemption and reconciliation, starting over, and the perfect society that could be built from a tabula rasa. To paraphrase the words of Rwandan writer Louise Umutoni in a 2016 interview, the genocide is important, but it is often told as the only plot point in a single story that we (non-Rwandans) tell ourselves about Rwanda.

I want to be careful here. I am in no way seeking to downplay the traumatic violence of the 1994 genocide against the Tutsi. Nor do I wish to trivialize how the genocide continues to be a source of trauma experienced by those who must live with its residue. That trauma very much continues into the present and is part of ongoing debates about political and racial subjectivity in Rwanda (see Burnett 2012; Nsabimana 2023; Hudani 2024). But that reality is not referenced when Kigali's postconflict status is evoked as a utopian plot point in building a new city for the end of the world.

For example, in a 2007 online article that covers OZ's work in Rwanda, a US journalist picks up on this plot point as it is told to him by a US architect who worked on the plan with OZ. "'The solutions we are looking at [in Kigali],'" the architect tells the journalist, "are just so much better than what we're able to do in the States'" (Blum 2007). The journalist concurred. Postgenocide Kigali, he wrote, "is an urban planner's dream: to draw lines on a map and shape the ideal city."

An excellent reporter, the journalist then rhetorically asks the question that must be asked about every utopian project: "a city for who?"

Utopia is a concept that gets thrown around a lot in urban studies. In scholarship on twentieth-century modernist cities, it is used to describe ambitious plans to build new societies through the reorganization of urban space as a way of sidestepping political and social revolution (Holston 1989, 57), or an authoritarian's dream of a perfectly ordered city (Scott 1998, 114). Scholarship on twenty-first-century planning in Kigali tends to use utopia this way—describing the plan as a "modernist" project in which the object is a "state utopia," a "documented utopia: a combination of fantasy and futurity" (Hudani 2024, 130). The plan, in this thinking, is a government attempt to convert a "post-conflict dystopia to a heroic wellspring of developmental success" (Hudani 2024, 196).

While this conception of utopia-as-fantasy makes sense in a certain vein that treats utopia as hubris, hope, or folly, it also raises several questions about the formal properties and language used in plans and marketing materials. First, if the object of planning in Kigali is a state utopia in which government officials actually believe in a fiction that can be so easily debunked by outside analysts, why is it so effective with external audiences like the architect, the journalist, and (presumably) their audience described above? And if sustainable urbanism in Kigali is really a state project of nation building, why would the government of Rwanda hire only outside firms to do this work? As the Rwandan architect Christian Benimana recently pointed out to a US journalist, several Rwandan firms operating in Kigali have the technical capacity to do what foreign firms have done in the city (see Neuwirth 2021). Second, what is *dys*topian about Kigali's postconflict status? Boosters like the US architect in the article above clearly see and articulate Kigali's postgenocide status as an "opportunity" to build a "new city." The postconflict is an element that is hyped, and continually referenced, not presented as a barrier to development. The master plan *is* utopian, but not in the sense that it is unachievable, a fantasy or a folly. It is intentionally utopian, in the literary sense of the word.

In her review of utopia as genre, literary scholar Pamela Bedore (2017) identifies three conventions of the form: a setting in another place distinct from where the audience lives; a narrator who moves between the other world and the audience's world; and a society that is not perfect but seeks perfection. The first convention renders utopia a geographic concept. Utopia may be situated in another time—it might be futurity—but it is always another place, one distinct from the world of the audience but also, unlike fantasy, possible to reach and inhabit. The second convention, the subject position of the storyteller, sets up the narrator "as a liaison between the reader and the utopian community" (Bedore 2017, 7–8). As in Thomas More's sixteenth-century original, the narrator of utopia must have knowledge of both the audience's world (to know what is significant) and the utopia (the better world) to connect the two. Finally, utopian narratives offer a fundamentally flawed society that *seeks* perfection (Bedore 2017, 6; see also Kelley 2003).

Underlying and supporting Bedore's three conventions is the fundamental attribute of utopia-as-genre: the emotional weight that makes utopia both a useful marketing device and a tool of political revolt. The affective state that generates this emotional force is what the science fiction scholar Darko Suvin (1979) calls "cognitive estrangement." Cognitive estrangement arises when we encounter—as a reader or viewer—a world that is strange (the other world of utopia) but also recognizably plausible. As we learn to inhabit that other, strangely familiar world, we gain an alternative perspective on our own.

Utopia in urban planning also shares a problem with utopian literature. Whether in a city or in a story, utopia is static. Frictionless spaces, in literature or in cities, quickly run into narrative dead ends (see Merrifield 2000). But there is another way—the proto-utopian way—to produce cognitive estrangement without falling into the narrative limits of generically pure utopia: begin with the end. Generate cognitive estrangement by inviting audiences into an alternative world where, because the old world is destroyed, they can experiment with new forms without the usual consequences or difficulties of breaking things.

Not any end will do, however. To generate the required estrangement, the end must create a radical break with the past, one that definitively "separates what came before with what came after" (Berger 1999, 6). The end is what literary scholars call an eschaton—a clearing event that creates a break with the past and leaves behind an opportunity to build something new.

In the materials and reporting on sustainable urbanism in Kigali, the 1994 genocide against the Tutsi is the implied catastrophe that travels with the name *Kigali*, creating a unique opportunity to build what Worthington in the interview cited above calls a "new city" and an example of a "new Africa" as a greener, more ideal, better city.

In addition, as every critical utopian and every Africanfuturist writer knows, dystopia is not the antithesis of utopia; dystopia is utopia, told from the perspective of those who must pay the costs of the better world. Ms. Iradukunda identifies utopia from this standpoint in her narrative of the December 2013 floods. Evoking the same plot points that are used to sell Kigali—the end, after the end, and a new beginning—she tells Kigali's story from the perspective of a person who must live in and repair her life from the messes created by efforts to green the city as they interact with infrastructure-induced ecological crises.

THE POLITICS OF REPAIR

Kigali residents rebuilt their own city after the 1994 genocide against the Tutsi. The results were not perfect. Twenty-first-century Kigali emerged in conditions of postgenocide scarcity: scarce petrol, a crashed currency, scarce imports that were mostly cheap knockoffs and secondhand. Kigali's rebuilding reflected that reality. Its popular neighborhoods were low-budget operations. Roads were not great, and the city had its share of public health and security issues. The repaired city was also, by necessity, low carbon, based on pedestrian and popular transport, and centered around economies of recycling.

Mr. Habimana, a resident of Kigali who had lived in the city since 1979, was the first to call my attention to repair as a response not only to the destruction left behind by the genocide but also to the dispossession wrought by planning. Mr. Habimana was evicted from the first neighborhood to be demolished in the name of sustainable urbanism, Kiyovu cy'abakene, in 2008. Like some of his neighbors, he moved across the road and rebuilt his home and two rental properties in Kimicanga, where Ms. Iradukunda lived. When Kimicanga was demolished, Mr. Habimana moved across the wetlands to Kazaire (Camp Zaïre), where I met him. During one of our early conversations in his living room, Mr. Habimana grew impatient with my questions about loss and dispossession. He wanted to talk about what he had kept, not what he had lost. Standing abruptly, he walked over to his steel front door and put a finger on it. "From Kiyovu," he said (recorded interview, April 22, 2014). His metal roof and his steel windowpanes were from Kimicanga. He had rebuilt his home and rental properties from the materials he had salvaged from previous demolitions.

Through a lifetime of repair, recycling, and rebuilding, as well as everyday negotiations with low-level government authorities, Mr. Habimana, like so many other Kigali residents who live and work in the city, has acquired knowledge that would be considered specialized in other contexts: engineering, construction, maintenance, legal manipulation. This wide distribution of knowledge about how to assemble a built environment, make it work, and maintain it with limited resources is not unique to Kigali. Many of the popular modes of city building, repair, and recycling that feature in this book belong to what Patience Mususa, in her work on the Zambian Copperbelt, aptly calls *"trying"* in the double sense of making efforts to pursue uncertain results (trying) under difficult (trying) circumstances (2021, 13; emphasis original). A rich ethnographic record has captured how residents of cities as different as Kinshasa, Ibadan, Dakar, and Lomé live in the aftermath of structural adjustment and currency crashes (see de Boeck and Baloji 2016; Guyer 2004; Melly 2017; Piot 2010; Piot and Batema 2019; Simone 2004a) while inhabiting (with various levels of success) the ruins of colonial and independence-era modernist planning projects in Cape Town, Monrovia, and Tema (Makhulu 2015; Hoffman 2017; Chalfin 2023). In addition, scholars have focused on the politics of maintaining systems that keep cities such as Johannesburg and Dar es Salaam up and running under conditions of state neglect while engaging with the infrastructural residue of apartheid and neoliberalism (Degani 2022; Simone 2009; Von Schnitzler 2016).

Scholarship in repair studies shows that repair and the related activities of maintenance and recycling are more than a universal good, the "humble but vital processes" that are overlooked as background (Graham and Thrift 2007, 2).

Breaking things down into component parts and putting them back together are also political and ethical ways of innovating and responding to a world that is always falling apart (Jackson 2014, 221). Repair work reverses the processes of ruining and wasting (J. Doherty 2022, 6) by converting discards into useful things, while calling attention to the exploitative conditions that require repair in the first place (Corwin and Gidwani 2021).

The vital labor of repair, maintenance, and recycling is also about undoing—in Kigali, particularly, undoing the massive disarray generated by projects aimed at bringing sustainable "order" to the city while also addressing the ecological issues, the unequal distributions of environmental burdens and social and economic inequality, that are not being addressed by sustainable urbanism.

Engaging in unauthorized repairs to their built environments, Kigali residents also maintain municipal service providers and district-level security guards in ways that also maintain state authority (see Barnes 2017). But in the process, people can undo destruction with repair, ameliorate wasting with recycling, while hijacking the very agencies that are installed in Kigali to prevent what many in the city call *akajagari* (unsanctioned spaces and practices).

In her important work *Master Plans and Minor Acts*, Shakirah Hudani also focuses on the politics of repair in Kigali, but from the perspective of national state-building and postgenocide recovery. For Hudani, repairing the city from war and genocide occurs in two modes: master plans, or state-led interventions aimed at erasing the traces of the past, and what she calls "'minor acts' of repair and continuity, building a complex and organic peace [that is] alternatively a social process that contests state ordering through smaller-scale modes of reworking space, and alternative imaginaries of community" (2024, 18). Hudani's work is an important intervention in the scholarship on trauma and peace building in Rwanda that moves away from jurisprudence and toward the role of the built environment in postconflict recovery.

My focus on repair engages with Hudani's work but also takes repair in very different directions, extending repair beyond a metaphor for postgenocide reparations and into the ecological and material significance of working with the city while using the architectures and expressive cultures that come from repair. This book focuses on the collective work of repairing the city, not just through individual "minor" acts—a phrase that risks marginalizing repair while devaluing what are major sources of livelihood in the city—and not just after genocide, but through unauthorized collective acts of repair that happen in moments in which Hudani suggests repair is rendered impossible by state domination and a culture of compliance: after master plan–driven efforts to evict city residents and halt repair through zoning, laws, and coercion.

Mr. Habimana, for example, had a lot to say about the injustice (*akarenganye*) of displacement, but he also understood how to use repair against master plan projects. He and most of his neighbors refused to buy houses in Batsinda—a master plan project. So, the City of Kigali attempted to force them to purchase "sustainable" homes by destroying their existing neighborhood. Mr. Habimana could not stop the bulldozers from tearing his house down—and 126 families did end up in the Batsinda estate (see chapter 4)—but his knowledge of repair and recycling, as well as his knowledge of how the municipality and the city's ecologies worked, allowed him and most of his neighbors to quickly rebuild elsewhere, undoing the damage of demolition while using the city's resources to repair his life in the city.

Mayors and government spokespeople often voice exacerbation with noncompliant residents who rebuild popular spaces and economies outside of Master Plan zoning codes faster than city authorities can break them down (see Niyomwungeri 2017). City authorities connected to planning have tried (unsuccessfully) to ban housing repair and additions in places that are slated for demolition; they have outlawed (unsuccessfully) affordable bricks made with local clay; and they have tried to ban (unsuccessfully) imported secondhand fabrics that get used in near zero-waste fashion industries and support a network of urban market spaces. As I show in chapter 5, landowners like Mr. Habimana do not do this by "resisting" the state. The repair work of rebuilding necessarily uses the very institutions that are designed to prevent it from happening—community police units and local authorities—against other promotional offices built to support the Kigali City Master Plan.

The politics of removal and rebuilding, destruction and repair, wasting and recycling have relevance beyond Kigali. An ethos of producing new spaces by working with what is already there speaks to an ongoing and interdisciplinary debate around the political relevance of discards and relations of disposal in the Capitalocene (see Moore 2015; Arefin and Fredericks 2025). This debate has also animated media studies (Jackson 2014), anthropology (Tsing 2015), architecture (Stoner 2012), and postcolonial ecocriticism (Iheka 2021) and justifies the interdisciplinary methods in this book.

THE URBAN HUMANITIES

If our crises of ecology and economy are, as the literary critic and postcolonial theorist Amitav Ghosh (2016, 9) says, also a "crisis of the imagination," then we could benefit from perspectives on these crises that extend beyond technocratic explanations. And if the so-called urban crisis in the Global South is really a "crisis of urban theory because the form that the cities of the global south are taking, the economies that shape them, little resemble the normative developmental models and canonical case studies" (J. Doherty 2022, 5), then we could use a more creative approach to theorizing urban life in Africa.

Kigali: A New City for the End of the World seeks a unique approach to urban life in an era of capital-driven planetary crises by drawing on a diverse set of interdisciplinary methods, including ethnographic work from and about African cities;

ecocritical perspectives that bring environmental studies together with African literature studies; and Africanfuturism that engages with urban environments.

This is not the first book on an African city in the urban humanities. My own work is influenced by Ato Quayson's (2014) *Oxford Street, Accra*, which employs methods from literary studies to read the streets of Accra and from street ethnography to read literature; Kenda Mutongi's (2017) *Matatu*, which brings ethnography, history, and popular culture together in a study of popular transport innovation in Nairobi; and Garth Myers's (2011, 2016) works that draw (among a wide variety of sources) on ecocriticism to grapple with African urban environments. My own induction to this approach comes from seven years of working as an urban humanities scholar in an interdisciplinary department of African and African American studies.

By ethnography, I mean working with linguistic proficiency, long-term immersion in place, and learning by doing. From 2013 to 2015, I collected over five hundred unstructured recorded interviews with Kigali residents who were impacted by sustainable urbanism and a handful of unrecorded interviews with city planners and foreign consultants. I also participated in licit and illicit construction projects to learn how to work with the city's earth and wetland clay. As my research progressed, street traders and secondhand clothing dealers allowed me to shadow their work in the streets and behind the scenes to learn how the city's intricate systems of recycling and discounting are put together. In the ethnographic chapters of this book, I offer a brief explanation for how I know what I think I know—whom I worked with and how I gained access to each space. In addition to ethnography, I spent hours poring over planning documents, PowerPoint presentations, marketing materials, and UN-Habitat and government reports to learn how a diverse range of institutions had brought the Kigali City Master Plan into being. After my uninterrupted fieldwork ended, I returned to Kigali every year (usually twice a year) until 2023 to catch up with friends and interlocutors and to visit the places discussed in this book.

While ethnography can offer unmatched access to everyday life, African ecocriticism brings environmental studies together with postcolonial literary studies to challenge the assumption that "Africans lack the proper environmental sensibility and knowledge to take care of precious biodiversity hot spots and . . . that environmentalists' efforts in Africa need to be conceived and led by non-Africans" (Caminero-Santangelo 2014, 3; see also Nixon 2011). Ecocriticism originated in literary studies, but the perspectives from this approach are useful for studying twenty-first-century plans. As I show in chapter 3, sustainable plans for Kigali are not just plans; they are digital images, texts, and film, hosted on social media platforms, open ArcGIS databases, and government websites with narrative arcs and plot points. They require attention not just to their visual aesthetics but also to what ecocritical scholars identify as "language and formal features such as genre, plotting, and narration [that] operate to construct and deconstruct meaning" (Caminero-Santangelo and Myers 2011, 3). In addition, many ecocritical works provide unique perspectives on classic themes in African urban studies.

For example, in his book *African Ecomedia*, humanities scholar Cajatan Iheka (2021) identifies what he calls "imperfect media," popular media that ranges from Nollywood to South African lo-fi music recordings, often dismissed as amateur and deficient by critics. Iheka shows that, contrary to Hollywood's high-budget and high-carbon production processes, "imperfect media" is low impact, low carbon, often based on working with what is already there. The point, Iheka asserts, is not to celebrate poverty but to take seriously the proposition that imperfection offers an environmentally ethical alternative to high-carbon corporate media production (see Iheka 2021, 227).

Iheka's conception of the imperfect features throughout this book as an alternative to the reactionary green perfection in the Kigali City Master Plan and as a way of making these alternatives thinkable. I extend this conception beyond media production and into low-impact construction, repair, and recycling methods that are deemed deficient and unsustainable by expert consultants and city authorities.

Arguing for the value of the imperfect comes with the risk of being accused of celebrating scarcity and valorizing poverty. But it is through attention to repair, rebuilding, and recycling that the causal processes of scarcity and poverty—global environmental racism, new strategies of accumulation, wasting—come into stark relief. We can and should be critical of the conditions that force people to innovate with their built environments and to recycle discards from other places. However, to simply ignore the expert labor that goes into making the city livable under conditions of demolition and displacement would be to—as other scholars who write about Kigali have done—reproduce the discourses that justify demolition and displacement, reducing Kigali residents to silent, non-agentive subjects who lack the vision and capacity to be stewards of their own urban environments. As I show in chapters 4, 5, and 6, backed up by very public debates about compensation in Kigali, city managers also recognize the value of the technologies, economies, and knowledge that go into making the popular city and use the discourse of deficiency and lack to plunder that value without fairly compensating Kigali residents.

In other words, the urban humanities approach in this book is more than an academic adventure into multiple fields. It is the necessary perspective to grapple with how green capitalism in an era of capital-induced climate crises makes and breaks cities of the future while accurately demonstrating who bears the cost for producing that future.

While the term *Africanfuturism* is relatively new, coined by the author Nnedi Okorafor in a 2019 blog post, there is nothing new about science fiction from Africa. A burgeoning scholarship on Africanfuturism as expressive culture in postcolonial literature engages with climate crises, political theory, and utopia from the perspective of people (Africans) and places (Africa) that have been erased from conventional renderings of the future (see Adejunmobi 2016; Cleveland 2024; Moonsamy 2016). And, as Amitav Ghosh (2016) compellingly argues in his work on the climate crisis, humanity's current predicament is sometimes rendered thinkable only through the estranged position of speculative fiction.

This book is not about Africanfuturism (see Cleveland 2024 for an excellent overview). I only draw on a handful of short stories, novels, films, and visual media that have been catalogued as Africanfuturist, and I read these texts for their theory rather than as objects of analysis. My intention is to show how African sci-fi offers insight into many of the themes of twenty-first-century African urban studies—utopia, climate crises, green capitalism—from the estranged position of alternative futures in ways that offer fresh perspectives on the city. I choose these citations, often over established and already well-cited academic voices, to show that African artists are also grappling with the same issues in productive ways. For instance, the Kinyarwanda/Kirundi cyberpunk musical *Neptune Frost* (discussed in chapters 5 and 6) engages with (among many other things) recycling as a way of calling attention to the global extraction and dumping industries and the vital labor of those who do the work of reversing the wasting processes. Tendai Huchu's short story "The Sale" (discussed in chapter 3) teaches us how to read plans for investment-driven sustainable urbanism by bringing to life a displaced future in which privatized sustainable urbanism is successful; Tlotlo Tsamaase's short horror story "Behind Our Irises" (discussed in chapter 4) illustrates how dispossession of knowledge works; and Lauren Beukes's biopunk classic *Zoo City* (discussed in chapter 5) offers a theory of "punk urbanism" that compellingly makes sense of unauthorized repair and the amorphous and ubiquitous term that populates so much popular and academic discourse on Kigali: *akajagari*.

A brief explanation of pseudonyms

All names used in this book, except those already published in media accounts and public documents, are pseudonyms to protect the identities of people I have quoted. To avoid manufacturing the familiarity that is conveyed with common anglicized first names, I use English salutations (Mr. and Ms.) with pseudonyms that are very common Kinyarwanda names (the equivalent of Smith or Jones). Other features—physical features, exact locations of residences—that might identify individuals have been omitted. None of the pseudonyms I use are unique, and they cannot be mistaken for another person. For example, the person I call "Ms. Iradukunda" above is not named Iradukunda, and she is not related or connected to any of the many people in Rwanda and around the world who are named Iradukunda.

A NOTE ON THE POPULAR

Most of this book takes place in what I call the popular city, and what in Kinyarwanda are called *katziye* (neighborhoods, from the French *quartiers*) and sometimes referred to as *akajagari*—neighborhoods built by the people who live in them and rendered unauthorized by building and zoning codes. These neighborhoods make up somewhere between 60 and 70 percent of Kigali's built environment (depending on who is asked) and house an estimated 79 percent of its population (see Bafoe et al. 2020, 22). The popular city encircles Kigali's original colonial core and then is patchworked through the city's thirty-seven wetlands and up its steep slopes. It is diverse in building typologies and in people. Plenty of people who identify as poor rent there, but so do large numbers of middle-class landowners, entrepreneurs, and professionals.

I use the term *popular* not as a politically correct substitute for *slum* or to sugarcoat urban poverty, but because the term is accurate. In my research, no one whom I spoke with used the word *slum* to describe where they lived. In Kigali, *slum* is used by city officials and some media, usually to justify demolition. Popular economies and spaces are popular in the sense that they are run by the majority and are crucial to maintaining the city (see Guyer 2004; Mutongi 2017 for similar uses of the popular). And rather than being inhabited by a homogeneous "urban poor" invoked by the concepts of informal settlement and slum, Kigali's popular neighborhoods are diverse and mixed use. They are not slums but neighborhoods: places with names and histories built by people who know how to envision and enact their own urban futures.

Some, for very good reason, have suggested that suspending concepts like "slum" is mere "polite, managerial language" that depoliticizes the harsh realities of poverty (J. Doherty 2022, 14). But as I show in chapters 3 and 4, *slum* and *informal settlement* (two words that are used interchangeably by planning firms and UN-Habitat) are the managerial discourses of sustainable urbanism. The language of slums and informal settlements is a well-known tool used by urban authorities to catalogue neighborhoods as *un*sustainable, cheapening the lives of people who live there and lowering the compensation costs of dispossession and displacement. Separating the planning discourse of *slums* and *informal settlements* from how Kigali residents conceive of their own neighborhoods is useful to track how these processes work.

WHAT COMES AFTER

The chapters that follow are organized along the historical production (chapter 2), branding (chapter 3), destruction (chapter 4), repair (chapter 5), and recycling (chapter 6) of urban space in Kigali. The book is organized both thematically and (more or less) chronologically, with chapters 2, 3, and 4 focusing on planning and destruction, while chapters 5 and 6 address popular politics of unauthorized repair, maintenance, and recycling. Each chapter is written on a different scale and from the perspectives of a different space and different people as they navigate sustainable urbanism in Kigali. Chapters 3 and 4 focus mostly on the logics of city branding, planning, and design, while chapters 5 and 6 are written from different spaces of repair and recycling to show how people navigate sustainable urbanism in the streets.

Chapter 2 is about what came before the Kigali City Master Plan. It follows the histories of colonial-era science fictions about a lost race in central Africa, land transfers, and technologies of city and urban authority building that gave way to the metropolis that is Kigali today. It is about the histories of clay architectures and urban authorities. These histories of seemingly mundane things like land transactions and clay architectures are important because they establish that what twenty-first-century planners call Kigali's "informal settlements" are not random, but places that have been built intentionally over time through systematized practices. Chapter 3 picks up in the twenty-first century. This chapter charts the origins of the Kigali City Master Plan as a distinct process that is built on branding Kigali as a utopian investment opportunity. It also outlines the contradictions of sustainable urbanism and shows how the city's brand image disrupts the everyday processes of city management as authorities come to be ruled by the fictions of their own creation. Chapter 4 is about imperfect architectures and making new markets for green commodities through the production of scarcity and the dispossession of clay building technologies. The focal point of this chapter is an unassuming artifact: a two-bedroom house designed to solve a "housing crisis" of "slums" that existed only in the master plan and UN reports.

The focus of chapter 5 is Kigali's unauthorized architectures and economies of repair and maintenance that rebuild and maintain the popular city after master plan destruction. This chapter draws on African cyberpunk and the "punk" technologies of unauthorized repair and street economies that supported it. Chapter 6 is about the recycling of fashion and spaces both in Kigali's popular recycling economies and in divergent agendas to use the plan to revalorize wasted space.

The book concludes in the epilogue by drawing on the tensions between the first chapters of the book, a critique of top-down sustainable urbanism and the fictions that support it, and the later chapters, which offer an ethnographic account of how Kigali residents navigate green capitalism. In this epilogue, I return to the Africanfuturist works—short stories, novellas, films—that make possible the book's analyses of utopia, dystopia, and alternatives to sustainable urbanism. Bringing these texts from the background to the foreground, I make provisional suggestions for an urban theory from Africa through an urban humanities perspective.

A FINAL HEDGE

This book is a critique of market-driven sustainable urbanism, not a cultural critique of Rwanda, the Rwandan state, or the ruling party. Readers familiar with the specialized field of Rwanda studies may be surprised or even troubled by how this book deals with the ruling party of Rwanda as neither a benevolent miracle maker nor an exclusively violent regime. Executive state authority certainly plays a role in hiring outside consultancies and creating the conditions for Kigali to be a center of green capitalism. But very few people in Kigali interact with the state at the level of executive authority. Most people in Kigali negotiate and contest with lower-level institutions that are much more malleable.

In addition, this book aims to challenge the tendency to treat Kigali as a case study of national state building while reducing the people who live there to an "urban majority traumatized by genocidal conflict, cowed by restrictive legislation, and with little history of unionism or activism to build on . . . " (Goodfellow 2022, 271). It is as though the pertinent academic question is whether we (outside scholars) should vote for the RPF.

Kigali does address the role of heterogeneous state institutions in facilitating and blocking efforts to remake Kigali as a new city for the end of the world. Because the Kigali City Master Plan is not the state but an instrument that is leveraged and struggled over by many actors, it offers privileged insight into how those dynamics work (and how they break down). Rather than producing yet another case study of state violence, I attend to how people build and repair their own urban environments in ways that are environmentally significant, contest the destructive process of green capitalism, and (I believe) offer all of us lessons for how we might live on a broken planet.

2

Production

Making an African Metropolis

On Saturday, July 27, 2013, I hiked up Mount Kigali by foot from where I was staying in Nyamirambo to a neighborhood that was then above the city's service grid. A week earlier, an *umuyobozi w'umudugudu* (community leader) invited me there to participate in *umuganda*, monthly compulsory community service, as a way of introducing myself and my research.

The road that winds up the face of Mount Kigali from Nyamirambo offers excellent views of the booming metropolis below, where close to two million people live on the slopes of three mountains and several smaller hills and around the thirty-seven interconnected wetlands that transect the city. Near the summit, you can see the city's new skyline of US hotel chains, German banks, and Finnish hedge fund–owned shopping centers. Pushing right up to the gates and walls of affluence are neighborhoods where the majority (60–79 percent, depending on whom you ask) live: the iron roofs of dense, single-story houses, bars, clubs, cafés, gyms, and shops, shaped by three generations of people who do the essential labor of maintaining the city.

On that umuganda day in 2013, my job was to help carry poles cut from trees in the nearby forest to a building site. At the construction site, about fifteen people worked, stamping earth into a foundation with hoes and digging holes for posts with shovels. About ten more people stood around watching, taking breaks to comment and joke about our progress. Once I dropped a pole off at the site, two people picked it up and sank it into holes dug in the stamped earth. Bamboo (African alpine)—purchased just outside Kimisagara Market below the neighborhood—was then woven horizontally through the vertical poles to create a frame. Then, two *abafundi*, masons skilled in construction trades, directed dozens of people to slap a mixture of clay—harvested from several feet below the topsoil in a pit at the building site and mixed with water, earth, and vegetation—onto the walls. By the noontime end of umuganda, we had completed a rectangular structure of four load-bearing walls a little over three meters high. It had no roof yet. The earth needed to dry; another layer would be added the following day, and another the day after that. The wattle and daub would later be replaced with air-dried bricks, walls would go down, additions would be made. A year later, the structure was connected to electricity. The house, like the neighborhood it was in, the city it belonged to, and the multiple authorities that governed it, would forever be a work in progress.

In Rwanda, every umuganda day ends with a meeting in which umudugudu (plural *imidugudu*) leaders communicate directives from above. In Kigali, one of the most frequent themes in umuganda meetings is *akajagari*, unauthorized construction, repair, and commerce. On that day in 2013, the umudugudu leader warned against building, making additions to, or repairing any structures raised without an official permit from the sector office, where each building plan is reviewed for compliance with the master plan.

During the meeting, no one objected that we had, under the direction of community leaders whose job it was to report unauthorized construction and repairs to cell authorities who supervise imidugudu, started the frame of a house without a permit, using materials and methods that were at that time technically banned (clay wattle and daub) on a plot that had been created outside of the official titling laws—also technically out of code. The entire neighborhood near the summit of Mount Kigali, like all its adjacent neighborhoods, was technically illicit: unpermitted, densely populated, in an area that was zoned for ecotourism. The construction proceeded as if there were no contradiction between the two programs: the master plan, which prohibited the neighborhood from being there, and umuganda, a program that predated the master plan and was used to improve access to housing and infrastructure through community service while enforcing fidelity to the ruling party.

In later interviews with people who participated in umuganda that day, everyone agreed that the house was noncompliant with the master plan. But no one saw the construction of the house as an act of resistance to urban authorities. Every step of the construction process, from the creation of the plot to the mobilization of community labor through umuganda, reified other aspects of ruling party authority against urban planning. By ignoring one state project—zoning codes in the plan—and by participating in umuganda, we extended the reach of the government into the city's edge through the provision of illicit housing and services. After all, many reminded me with an often-cited slogan, "Aren't the people the state [*abaturage si leta*]?" That day, the very institutions that were designed to prevent illicit housing through decentralized surveillance (umuganda, local leadership) were redirected to facilitate the construction of the house and legitimate its existence.

The following chapters are about sustainable urbanism and how Kigali residents navigate demolition and displacement through unauthorized repair and recycling practices. This chapter situates the contemporary politics of city building in Kigali within a hundred-year history of what David Morton, in his work on Maputo's concrete city, calls "a politics of housing and infrastructure that did not always call itself politics" (2019, 11). Unlike the concrete urbanism in Morton's work, Kigali was (and to a large extent still is) built with earth and sand largely secreted by the wetland ecosystems and supported by legal systems that predate and are often alternative to contemporary authority in the city. This chapter also serves as a corrective to other recent work on Kigali that elides this material and spatial history of the city in the service of treating planning in Kigali as a case study of postgenocide national state building. Putting the elements that went into constructing an out-of-code house and making it legitimate—clay architecture, alternative land transfers, lower-level authorities—into a historical view illustrates how today's struggles over the means of building and repairing the city belong to a continuation of earlier practices of construction, repair, and the reification of state power.

Those processes began during colonialism, which was certainly not a master plan but a series of "sometimes absurd episodes" in the late nineteenth and early twentieth centuries that involved private explorers on adventure, a literature of fictions about a lost race in central Africa, and ecological crises from which a tiny fort called Kigali emerged (Chrétien 2003, 218–20). As peripheral as Kigali was to other colonial settlements in the region, colonialism nevertheless had an impact on Kigali and its contemporary form and efforts to racialize the city through spatial hierarchies. After formal independence in 1962, Kigali's first generation of independence-era residents employed colonial-era hybrid clay architecture and innovative land transfers to partially decolonize their own city and, in the process, built out state institutions. These practices set Kigali up to become the center of postgenocide repair at the turn of the twenty-first century, more than a decade before consultants with OZ Architecture claimed to be "rebuilding" the city.

COLONIAL FRONTIERS AND SCIENCE FICTIONS

There are Rwandan oral traditions about the late nineteenth-century expansion of empire into the region and its effects on Rwanda's ecological and political landscape. These accounts begin with an omen of the catastrophes to come: a solar eclipse (*ubwrira kabiri*, the second night) that happened on December 22, 1889 (Botte 1985, 53). What followed was "an apocalyptic surge" of diseases, famines, and the destruction of livestock. No one is sure which of the microscopic invaders that traveled to East Africa from Europe and South America via global trade in the region killed the Rwandan Mwami (King) Rwabugiri, who, until that point, had kept enslavers, missionaries, and explorers at bay (Newbury 1988, 40). But in 1895, Rwabugiri died suddenly on a boat crossing into what is now the Congo (see Vansina 2004, 177).

A year after Rwabugiri's death, his chosen successor, Mwami Rutarindwa, was assassinated in a coup, throwing the region into political chaos as factions vied for authority. Rutarindwa's adoptive mother, Kanjogera, defied a century of the rules of Mwami succession and installed her teenage son, Mwami Yuhi V Musinga, in power (Vansina 2004, 178). In 1897, with no real political legitimacy and facing threats to their authority throughout the region, Musinga's family entered into a partnership with the German military, which had been trying to establish a presence in Rwanda for several years and had been waging its own reign of terror and mass murder in what is now southwestern Rwanda (Louis 1963, 122; Newbury 1988, 54). A year later, a German military expedition brought the German psychiatrist and explorer Richard Kandt into the Kingdom of Rwanda.

In 1904 Kandt achieved some celebrity in Germany with his book about his travels in central Africa during the late 1890s. The book's Latin title, *Caput Nili*, plays on the phrase's double meaning, the search for the Nile as an impossible pursuit and Kandt's search for a source of the Nile. In *Caput*, what Kandt did report back about was what he described as a lost civilization made up of three distinct "peoples" (*volk*). He encounters "a population of Bantu Negroes who are called the Wahutu, a different volk from Watussi who are either of Semitic or Hamitic origins . . . " (Kandt 1904, 257). The Watussi, Kandt explains, "tower at over two meters tall and recall the worlds of fairy tales and legends" (1904, 258). Finally, Kandt describes the "Batwa," a "tribe of dwarfs who live in the caves and marshes" (1904, 259).

To be sure, Kandt's *Caput Nili* is an example of what scholars of both Euro-American and African science fiction have identified as the colonial-era travelogue roots of the genre (Adejunmobi 2016; Rieder 2008). It is a story about a European scientist on adventure in central Africa who encounters a lost civilization (the Tutsi) that rules over a distinct "race" (the Hutu), with a mysterious tribe of "dwarves" (the Twa) on the margins, with no reference to the violence and ecological destruction of the moment that features in Rwandan oral traditions. But while the late nineteenth-century colonial project laid the foundations of science fiction as a distinct genre, its prototype, the colonial travelogue, like Kandt's *Caput*, was also crucial for selling colonialism to audiences back home while filtering African civilizations through European constructions of race.

Back in Germany, Kandt's travel across space to Africa helped popularize the German colonial project as an adventure into a world "untouched by time and foreign elements" that he could "lay out for science" (Kandt 1904, 258). In his account of Rwandan society, Kandt confirms the Hamitic hypothesis, an expression of (now debunked) nineteenth-century European race science—the lens through which nineteenth- and twentieth-century explorers and colonial administrators saw their own constructions of race at work, in Rwanda and elsewhere, as biological fact.

In 1906 Kandt was appointed to represent colonial power in Rwanda as German East Africa's first advisor to the mwami (king), known as the resident. "The selection of the first Resident," wrote the diplomatic historian William Louis in the 1960s, "offered no problem. Was there anyone more qualified ... than explorer, poet, and scientist of Rwanda, Dr. Richard Kandt?" (1963, 146).

THE FOUNDING OF KIGALI

Rwanda had centers of administration, political power, and trade centuries before the expansion of European empire in central Africa. These capitals moved with the mwami, the mwami's mother, and their administrations around their always shifting territories as they expanded politically westward and economically eastward (see Vansina 2004, 63, on ambulatory capitals; Newbury 1988, 23–54, on territorial expansion). According to one of his biographers, Kandt considered this mobile authority a problem. He "could not follow a 'Sultan' who was always on the move" (Bindseil 1988, 108). The German administration decided to contain the mwami's palace in the town of Nyanza and build a new center of colonial administration on a sparsely populated hill 91.5 kilometers north of Nyanza, close to the geographic center of Rwanda and along a preestablished caravan trade route (see Bindseil 1988, 108; Sirven 1984, 135).

On September 14, 1908, Kandt wrote to the colonial governor of Dar es Salaam, asking for the fort to be recognized. But from its naming on, Kandt did not have total control over his settlement. "When I founded the Residence," he wrote, "I named it after the hill that it is on, Nyarugenge." But Rwandans, beginning a tradition of frustrating state authority by hijacking naming processes that continues to this day, chose a different name. "The name Nyarugenge," Kandt complained in his letter, "absolutely fails to be taken up. On the contrary—and I could cite a hundred examples—the natives, from the Sultan to the last peasant, call the station Kigali [after the larger mountain nearby]. . . ." Kandt closes the letter by "humbly asking permission to name the seat of the residence, Kigali" (Bindseil 1988, 108).

The permission was granted. Kigali was officially recorded in Dar es Salaam as the headquarters of the Residence of Rwanda on October 15, 1908.

Unsurprisingly, colonial-era Kigali gets scant mention in histories of Kigali and less in histories of Rwanda. What Kandt called the *boma* (fort) was only a tiny outpost that served other German forts in the region. There was no plan for the boma to become the booming metropolis it is today. Unlike the British towns of Kampala, Nairobi, and Zanzibar, Kigali had no travelling technocrats like W. J. Simpson to experiment with sanitation and segregation, or to make it a node of empire. Unlike in West Africa, there was no Maxwell Fry or Jane Drew to innovate tropical architecture. The boma was an extreme example of what William Bissell

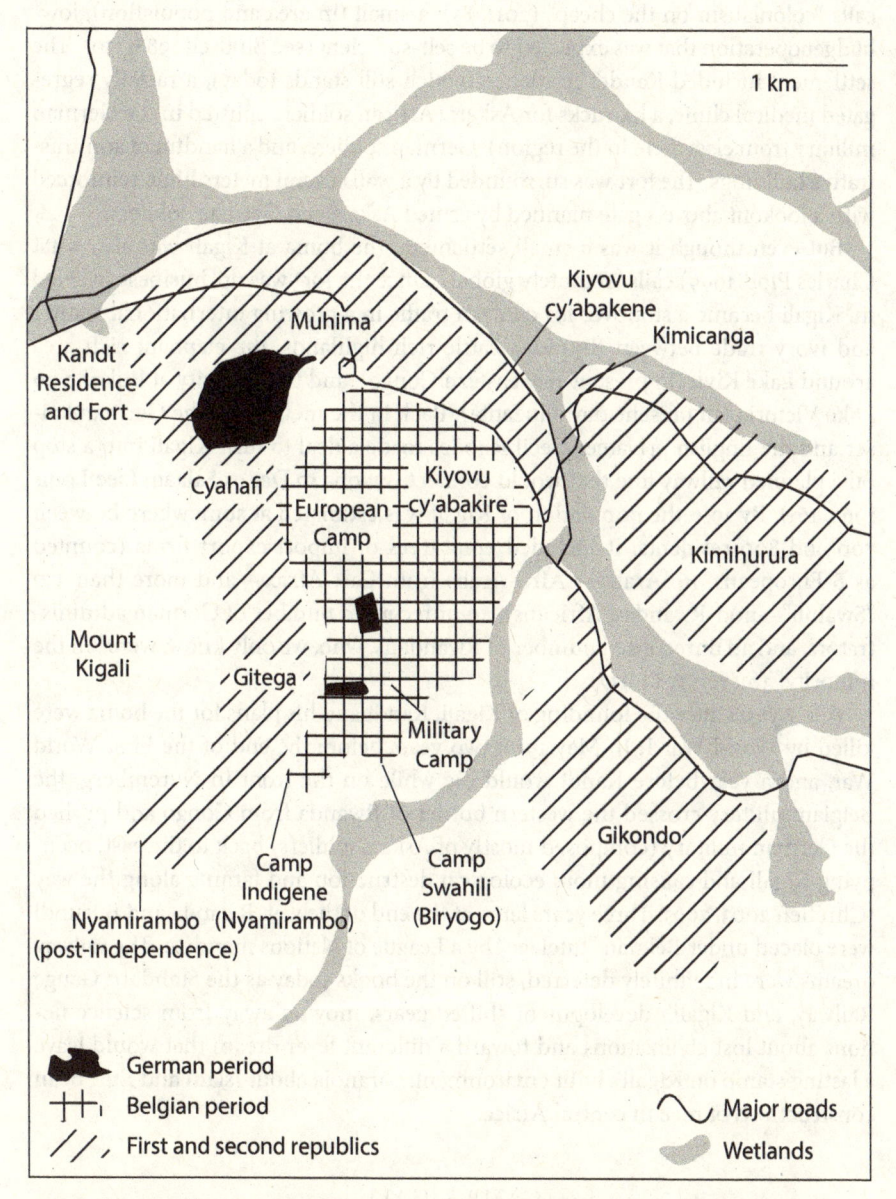

MAP 2. The growth of central Kigali on the top of Nyarugenge Hill and across the wetlands from 1907 to 1916 (the German Period), 1916 to 1962 (the Belgian period), and 1962 to 1990 (the First and Second Republics). This map is made from multiple sources. It draws from Pierre Sirven's (1984, 136) historical map of the city and also the wetland buffers established by the City of Kigali (see https://masterplan2020.kigalicity.gov.rw/portal/apps/webappviewer/index.html?id=218a2e3088064fc6b13198b4304f3d35/).

calls "colonialism on the cheap" (2011, 85): a small (in area and population), low-budget operation that was expected to be self-sufficient (see Bindseil 1988, 130). The settlement included Kandt's residence (which still stands today), a racially segregated medical clinic, a barracks for Askari (African soldiers enlisted in the German military from elsewhere in the region), German soldiers, and a handful of administrative buildings. The fort was surrounded by a wall several meters high, reinforced with a lookout above a gate manned by armed Askari and German soldiers.

But even though it was a small settlement, the boma at Kigali was also what Charles Piot (1999) calls "remotely global." Once the fort was up, business trickled in. Kigali became a stopover for caravan traffic in a growing international leather and ivory trade between Rwanda's cattle-rich highlands, the elephant-rich area around Lake Kivu (in present-day eastern Congo), and the port city at Bukoba on Lake Victoria (in present-day Tanzania). Back in the metropole, the German kaiser and the English privateer Cecil Rhodes made a deal to turn Kigali into a stop on a planned railway line that would connect Bukoba to Dar es Salaam (see Louis 1963, 161). By 1916 the population of Kigali was estimated at somewhere between 700 and 800 residents. It included employees of import-export firms (counted as 6 Europeans, 16 "Arabs"—Afro-Arabs from East Africa—and more than 350 "Swahili"—non-Rwandan Africans), an unrecorded number of German administrators, and an unrecorded number of Rwandans, who, we only know, were "in the minority" (Sirven 1984, 137).

A few years after the founding of Kigali, Kandt and his plans for the boma were killed by World War I. In May 1916, two years before the end of the First World War, and a year before Kandt would die while on the front in Nuremberg, the Belgian military crossed the western border of Rwanda from Congo and pushed the German military (composed mostly of Askari soldiers) back to the east, occupying Kigali and causing more ecological destruction and famine along the way (Chrétien 2003, 260). Three years later, at the end of the war, Rwanda and Burundi were placed under Belgian "tutelage" by a League of Nations mandate. The railway dreams were indefinitely deferred, still on the books today as the Standard Gauge Railway, and Kigali's development shifted gears, moving away from science fictions about lost civilizations and toward a different fever dream that would leave a lasting stamp on Kigali's built environment: paranoia about Islam and European constructions of race in central Africa.

CAMP KIGALI

The Germans wanted to open Kigali to trade through railroads while governing rural Rwanda through the mwami. The Belgians wanted to shut the fort down and enclose it and break the mwami's power. The Belgian colonial administration expanded its rural business interests into Rwanda and continued to support (and profit from) the institutions of indirect rule—specifically, forced labor supplied by land chiefs and justified as a "traditional" form of "feudalism" (see Newbury 1988). But the Belgian administration disagreed with Kandt's choice of Nyarugenge Hill as an administrative center and moved the colonial capital of Ruanda-Urundi to Usumbura (present-day Bujumbura in Burundi). It then moved the administrative capital of Rwanda from Kigali to Astrida (present-day Huye), relegating Kigali to the status of a slow-growing colonial backwater.

The Belgian administration also had different ideas about the small but cosmopolitan community of Asian, Arab, and African merchants who had taken up residence in Kigali and made up the majority of its population. Kandt believed in the potential to develop revenue by taxing the regional caravan trade and encouraged Baluchi, Omani, and Afro-Arab merchants to settle in Kigali (Louis 1963, 169). The Belgian administration saw Indians and Omani as potential English agents, while Islam was seen as a political threat to the business of European plantations and the missions in the region. "The largest risk, in the eyes of Belgians," wrote Rwandan historian José Kagabo, "was that Muslims would teach 'the Blacks' to hate 'the Whites'" (1982, 36; capitalization and quotation marks original). To contain these apparent threats to colonial power and to the Catholic missions in Rwanda, Belgian colonial authorities replanned Kigali, and other towns in Rwanda, into a series of camps (their word) that would, in theory, keep Rwanda out of town life and Rwandans out of the towns.

The identity *Swahili* has always been fluid, indicating a hybrid of African, Arab, and South Asian cultures. But the Belgian colonial administration in Rwanda used the term as an umbrella category for any African Muslim (Rwandan or non-Rwandan), thereby classifying all African Muslims—the Askari and their descendants and Rwandan converts and sometimes Omani and South Asian merchants—into a single, fixed category. They then stamped that category onto Kigali's built environment, using the production of space to produce racial categories.

The first Swahili Camp in Kigali was built in 1928 where École Technique Muhazi stands today, just a couple hundred meters east of the Military Camp. The Belgian colonial authorities also built the European Camp, just north of the military barracks. Then, in 1937, citing overcrowding, Belgian colonial authorities demolished the original Swahili Camp on the northeastern edge of Kigali and relocated the town's "Swahili" population to an area behind the military camp in Nyamirambo, where Biryogo is today (recorded interview with former resident of Swahili Camp, October 29, 2014; supported by secondary sources—see Sirven 1984, 502). Belgian authorities created another camp in 1951 for Rwandan *évolués*, French-speaking Rwandan civil servants, just south of the Swahili Camp in what is present-day Tapis Rouge (in Nyamirambo).

By the eve of independence, Kigali was a small collection of segregated but porous camps packed on the top of Nyarugenge Hill, surrounded by wetlands: the Swahili and indigenous camps were on the south side of Nyarugenge Hill, and the European Camp was on the north side, with the military camp in between, serving as a buffer (see map 2). Kigali was segregated internally but also segregated from the countryside. Armed toll stations collected a heavy entrance tax that was used to keep rural Rwandans out of town.

Of course, many Rwandans, incentivized to escape corvée labor obligations on rural plantations and road-building projects, accessed town and contributed to its growth. As one former Swahili Camp resident told me, "You could sneak into town, put on a kaftan, and then . . . you were Muslim! [laughter]" (recorded interview, October 29, 2014). Still, even this declaration of fluidity also highlights colonial aspirations to demarcate camp from town, and foreigner from Rwandan, by racializing the categories of Swahili, indigenous, and European in space. If you were Rwandan and you wanted to enter town, you needed to play the part of a "Swahili" and stay south of the barracks at the Military Camp, convert to Islam and speak Kiswahili, or be a French-speaking *évolué* "indigenous" elite with a Catholic mission education.

Kandt's fort and the Belgian camps were slow-growing, low-budget, incomplete works in progress directed by nineteenth- and twentieth-century European constructions of race and paranoia about Islam. In many ways, the processes that built colonial Kigali echo the development of some other colonial cities in other east and central African contexts. It was driven "less by scientific rationalities" and more by European constructions of race and paranoia (see Bissell 2011, 214; Fabian 2000). But the establishment of the fort and subsequent camps based on these conflicting ideas nevertheless created the partially segregated base of Kigali's built environment that still stands today: the European Camp is today the central business district and the beginning of the elite enclave of Kiyovu (where many Europeans and elite Rwandans still live); the Swahili Camp became the thriving working-class (and still predominantly Muslim) neighborhood of Biryogo; the indigenous camp is now Tapis Rouge in Nyamirambo; and the former military camp is Camp Kigali, which still stands as a barrier between Biryogo and Kiyovu (see map 2). In addition to laying the segregated base of the city, colonial (mis)management of Rwandan built environments and labor organization had other durable effects on Kigali's postcolonial form.

COLONIAL DURABILITIES AND HYBRID ARCHITECTURES

The one issue that historians, political scientists, anthropologists, and journalists tend to agree about when it comes to colonialism in Rwanda is that "colonialism helped lay a foundation of division and conflict in the country" (Purdeková and Mwambari 2021, 20). While Rwandan ethnic categories were not entirely colonial inventions, the legal and material infrastructures of colonialism worked to codify and produce racial identities. In the territory outside of Kigali, German and Belgian colonial authorities shored up the power of a minority class of "natural" rulers (the mwami and land chiefs) at the expense of a majority Hutu, non-elite Tutsi, and a minority Twa (see Burnett 2012, 43; Chrétien 2003; Lemarchand 1970; Mamdani 2001; Newbury 1988; Uvin 1998). Then, in a move that shifted anticolonial sentiment to anti-Tutsi and pro-Hutu ethnonationalism, the missions and colonial authorities became the champions of a new Rwandan elite who themselves claimed to be the champions of a class of dispossessed Hutu peasants (see Chrétien 2003, 286). But in addition to converting somewhat fluid social categories into fixed identities—a move that would have devastating effects on Rwanda—colonialism had other, unintended material effects on Kigali's future.

Colonial administrators and missionaries were as fascinated by Rwandan architecture as they were with social hierarchies and recorded and photographed "traditional" homes—records that now appear in archives and museums. There was much variation in form, size, and materials, but in general, late nineteenth-century architecture in Rwanda, from the mwami's palace to homes of subsistence farmers, were constructed with walls made of reeds woven together, alpine bamboo, and roofs thatched with grass. Generally, the roof was held up by internal wood pillars, and the frame was constructed of wood poles—usually African alpine bamboo—that rose from the ground and curved in to form a dome roof. The perpendicular posts were reenforced with horizontal poles that wrapped around the structure. Interior and exterior walls and thatched roofs were then sealed with reeds that were woven into the frame of the house, making the structure appear as a coherent whole (see Niyoyita and Li 2019). (See figure 1.)

Kandt needed to build his fort within the extreme limits of his budget and, as occurred in other colonial contexts, he did so in a way that would reinforce racial hierarchies that underpinned colonial power. As elsewhere in the region, the German fort and Belgian camps in Kigali differentiated racial-political subjects through built space. Kandt built his fort with clay that was cost effective—extracted from the wetlands below Nyarugenge Hill. Once "won" (harvested), the clay was then molded and baked into bricks and tiles that visually presented the boma in Kigali as a German, and later Belgian, center of power distinct from the mwami's palace (built with reed walls and grass-thatched roof) in Nyanza (see figure 2).

Kigali's small population of Swahili, Omani, Afro-Arabs, Indians, and Europeans were certainly not the first people to work with clay as a building material in Rwanda. High-grade clay was then, and still is, readily available in the region. It was also used as a sealant for reed structures and belongs to a long cultural history of pottery. But when the workers (mostly African corvée laborers) who constructed Kandt's fort built the first brick kiln in Rwanda, they installed a technology that would help shape the city's future built environment and would later be the source of struggles over its form (see Bindseil 1988, 110).

After World War I, missionaries and European entrepreneurs began setting up their own brick kilns throughout Rwanda to build missions and churches and to

FIGURE 1. Reconstruction of nineteenth-century Rwandan reed home with thatched roof at the King's Palace Museum in Nyanza. Photo by Samuel Shearer, 2015.

FIGURE 2. Postcard of Kigali's military camp (where Camp Kigali, in the center of Nyarugenge, is today) during the Belgian occupation, stamped 1916. Note the walls and posts made with baked clay bricks and clay tiled roofs, all made on-site. Source: Samuel Shearer, personal collection.

export Rwandan bricks to other markets (Newbury 1988, 175; Jefremovas 2002, 21). Belgian missions, sometimes in collaboration with colonial authorities, also engaged in a few experiments with brick housing models designed to transform the shape of the built environment and mold Rwandans into proper colonial subjects. For example, in her work on the history of colonial and postcolonial household architecture in Rwanda, Jennifer Gaugler documents how prototype homes built with baked bricks instead of reeds were erected in rural areas and beta tested by missionaries who believed in "a change in building materials, to make dwellings healthier and more comfortable" (Gaugler 2018, 43). Gaugler argues that these projects (as in so many colonial contexts) were about not just improving housing but also producing colonial subjects. Colonial and mission authorities believed that homes built with bricks "might also persuade the indigenes to spend more time indoors, and so cultivate the kind of nuclear family life that seemed more appropriate by European standards" (Gaugler 2018, 43). But building with baked bricks (that were also in demand in export markets) turned out to be too expensive for all but the Rwandan elite (Gaugler 2018, 46). And besides, the missions and colonial businesses were more interested in exporting bricks to other markets. As Gaugler also shows, backed up by other architectural studies (see K. Smith et al. 2018), Rwandans did not reject clay outright. They incorporated the materials and shapes from colonial housing projects into their own hybrid architectures that simultaneously rejected European models of nuclear families and embraced the durable materiality and flexibility of clay.

Homes became rectangular rather than round; clusters of homesteads that connected extended families became buildings positioned around courtyards with a wide diversity of forms. While the process was certainly not linear, by early independence local clay and earth were taken up as core building materials in Rwanda, and artisanal brickmaking became "one of the most important sources of rural non-farm employment in the country" (see Jefremovas 2002, 21). Because outside consultants would later be credited for innovating air-dried clay construction in Kigali, it is worth detailing how clay gets extracted in Kigali and converted into building materials.

In her ethnography of brickmaking in 1980s Rwanda, Villia Jefremovas documents two general modes of winning clay. Just outside Kigali, as throughout Rwanda, clay could be harvested from the wetlands close to the surface with very little topsoil, or overburden, needing to be removed. In the 1980s, the locations of clay pits were chosen and governed by local (commune) authorities. At higher elevations (such as Mount Kigali), clay is dug out of a deep pit—usually on private property, often in powder form—and needs to be mixed with water (see Jefremovas 2002, 26). In the 1980s, brickmakers who owned artisanal kilns apparently controlled most of the harvesting process, while landowners would lease out pits at higher elevations.[1] While Jefremovas's work focuses on the production of

FIGURE 3. A common sight in Kigali, amatafari ya rukarakara drying before being used. Buildings in the background are made from air-dried bricks, sealed with plaster, with a base of baked bricks and a stone foundation, surrounded by a gate made partially with reeds and partially with iron. Photo by Samuel Shearer, 2022.

bricks and tiles baked in kilns prior to the genocide, other ways of working with clay were already established in Kigali.

By the 1960s, three general modes of constructing with clay were systematized into expert trades, often used together on the same projects. The most affordable and fastest method is wattle and daub (*ibiti na ibyumba*, literally trees and mud), described in the opening vignette to this chapter, in which earth, clay, and vegetation are mixed together for a building material. The most common method of clay construction in Kigali—at least today, and historically used by Rwandans with fewer resources—is construction with compressed air-dried bricks (*amatafari ya rukarakara*), often mixed with other materials. Then as now, amatafari ya rukarakara were significantly cheaper than baked bricks because they skip the kiln process, requiring less labor and access to resources. The compressed, rectangular bricks are formed by pressing prepared clay mixed with earth, and sometimes vegetation, into molds to remove as much oxygen as possible (which strengthens the clay), either by hand or with a tool press. Rather than being baked in kilns—a process that is labor intensive, carbon generating, and expensive—amatafari ya rukarakara dry in the sun, usually under a plastic tarp (to prevent damage), before they are used in construction projects (see figure 3). Like wattle and daub, walls built with air-dried bricks are sealed with *karabasasu*, a plaster made with coarse sand, sometimes mixed with a small amount of cement, lime, and often paint.

Finally, the most expensive method of working with clay, both in materials and in labor costs, are *amatafari yahiye* (burnt bricks), which, until the postgenocide period, could be purchased from artisanal kilns located just outside of Kigali.[2] Of course, building with clay and the construction techniques that rely on it are not unique to Kigali—and the argument here is not that Kigali is a unique "case" of working with these materials in the region. My point is that clay, and the wetlands that source it, played a unique role in shaping the politics of independence-era Kigali and postgenocide repair.

CITY OF CLAY

Kigali was named the capital of independent Rwanda on July 2, 1962. The town had a population of somewhere between five thousand (Sirven 1984, 138) and six thousand people (Jaganyi et al. 2018). Kigali had a landing strip, but no airport, and only two kilometers of tarmac roads. The president of Rwanda's First Republic was Grégoire Kayibanda, a pro-West, authoritarian leader of the "social revolution" that claimed to represent Hutu peasants. In 1964 and 1965, a team of French consultants who worked for the Cold War parastatal agency Secretariat des Missions d'Urbanism et d'Habitat (SMUH) developed a master plan to extend the city of Kigali beyond Nyarugenge Hill and create new housing in the south of the city. The plan envisioned migrant workers from rural areas renting low-income housing on the southern edge of the city in Nyamirambo, while the wealthy lived in exclusive enclaves on the north side of the city (see Sirven 1984, 519).

Kayibanda's government achieved only some of the first master plan for Kigali. The government constructed ministry buildings, but on the hill of Kacyiru rather than in Nyarugenge, where the plan envisioned them, and extended the cadastre of Kiyovu northward into Kimihurura, zoning the area for upscale housing for foreign consultants and the new Rwandan elite. Instead of building affordable housing, the Kayibanda government tried to restrict rural-to-urban migration with legal stratagems. An early effort was decree no. 4/9 of October 1, 1967, which prohibited the construction of any building within the city limits without prior authorization from the government (Sirven 1984, 531).

But the Kayibanda government had a problem that colonial authorities did not: Kigali was no longer a colonial outpost of camps. It was a capital city. It needed to convert its landing strip to an international airport, and it needed a network of tarmac roads connecting the landlocked capital to neighboring countries. With only a few thousand people, Kigali lacked a reserve army of labor to build the new city.

In the late 1960s, Mr. Gutsinda, who was originally from Rwanda's southern province, purchased a plot of land as close as he could get to the edge of Kigali's official town limits, just a few meters from the tollbooth that was originally used to keep Rwandans out of the European Camp in Kiyovu. "That time," he laughed, "there were more hyenas than people in Kiyovu."

"When did people start arriving?" I asked him.

"Later in the '60s, it started slowly, then it picked up, 1970s?" he guessed. " . . . They [contractors] started bringing all these workers from Cyangugu (the Rwanda-Zaire [present-day Congo] border) and the Congo . . . " (recorded interview, April 17, 2015).

Two World Bank reports from the 1970s, maps published by geographers working on-site at the time, and aerial photographs from the 1970s, as well as four other recorded interviews with former neighbors of Mr. Gutsinda who also arrived in the 1960s and 1970s, confirm Mr. Gutsinda's memory (World Bank 1972, 1979; Sirven 1984, 136; OZ Architecture 2007, 21). Building in Kiyovu cy'abakene began after colonialism in the late 1960s and picked up speed in the 1970s, after Astaldi-Estero, a former colonial contractor, won a $9.3 million World Bank contract to bridge the wetlands and lay tarmac from the town center to the new airport. With Kigali's labor shortage, the firm had to truck in its workforce from all sides of the Rwanda-Burundi-Congolese border. Astaldi-Estero was not alone. The Belgian Safricas-Impressa, the Italian Murri-Freres, and the French Entreprise Construction were all picking up World Bank, USAID, French, and Belgian contracts to build Kigali's roads, ministry buildings, and the national bank, and they all trucked their workers in from rural areas and surrounding countries.

Rwandans, Congolese, Burundians, and Ugandans arrived in Kigali to work on new construction projects. Instead of following the established colonial-era protocols in which migrant workers returned to their homelands after a project was completed, many of Astaldi's workers decided to stay in the new capital and ignore the 1968 building codes. They settled just north of the city limits, on the edge of the former European Camp, known then as Kiyovu cy'abakire/cy'abazungu ("Kiyovu for the rich/whites"). The families whose land went under Astaldi-Estero's bulldozers also relocated to the other side of the tarmac; their homes, as a popular Kinyarwanda folk song about this era goes, "were left to the termites."[3] Kigali's new workers then sent for their friends and families back home, and—in the absence of any municipal policy governing workers' housing—they built their own city on the edge of the European town and defiantly called it Kiyovu cy'abakene, "Kiyovu for the poor."[4]

To be sure, the wetlands just below Kiyovu cy'abakene that surrounded the colonial city limits were used as what the anthropologist Jacob Doherty in his 2022 work on Kampala calls a "sink" of sorts. The wetlands provided both a natural barrier to keep people out of Kigali during the colonial period and a place to deposit the spillover contradictions of the independence-era government's official anti-urbanist policies on the one hand and its need for cheap, disposable labor on the other.

But no one needed the government to hire an outside consultant to tell them that Kiyovu cy'abakene had its share of ecological and health issues. Mr. Nsengimana, who moved to Kiyovu cy'abakene to pursue a boxing career in the 1980s, put it this way: "At first, there were a lot of problems in Kiyovu . . . near the wetlands [*igishanga*] . . . . The mosquitos were a problem. And because of the way we dug toilets [laughter], we would have problems with cholera and other diseases" (recorded interview, February 8, 2015).

Being exposed to the environmental injustices of waterborne diseases and flooding was certainly the outcome of colonial planning that used the wetlands as a barrier to the center of town in efforts to prevent Rwandans from entering. And it was also the result of independence-era policies that cheapened labor costs by outsourcing the fixed capital of housing and services to workers themselves. But early independence-era Kigali residents were not only subjected to the harms of capitalism and the residues of colonialism, unable to act on their built environment. People who lived there understood these issues and worked to resolve them.

The wetlands were also a source of groundwater and clay to build new housing stock. Kimicanga (named after *umucanga*, sand) had rich deposits of clay needed to make amatafari ya rukarakara and sand for karabasasu as a sealant. Landowners like Mr. Gutsinda and Mr. Nsengimana's parents hired abafundi who used wetland clay to construct the first generation of houses with (mostly) amatafari ya rukarakara. As Mr. Nsengimana pointed out, once drains were built through a combination of community service and sector-level initiatives, "life [*imibereho*, livelihood] improved. Disease dropped, food was cheap, rent was affordable, it was close to everything . . ." (recorded interview, February 8, 2015).

While living near the wetlands was not ideal from a public health perspective, clay and groundwater also provided the materials for Kiyovu residents to do what the Kayibanda government refused to do: build an affordable stock of houses in the center of the city. In addition, Kiyovu cy'abakene's landlords disrupted the complicated land tenure processes with their own innovative reforms that—to this day—make it possible for those excluded from the city to gain access to its resources.

LAND, BEER, AND INDEPENDENCE

In early-independence Rwanda, laws governing land tenure were a mess of contradictory legal codes inherited from the colonial period that ranged from customary rights linked to lineage and patron-client relations to colonial strategies devised to alienate people from their land. Technically, the state owned all land in Rwanda, but land could be acquired through inheritance (*umunani*) and gifting. As Villia Jefremovas writes of this period, land was held through "a complex combination of various degrees of 'private ownership,' customary rights and government ownership" (Jefremovas 2002, 74–75). This muddle made getting access to land and the rights to build on it a complicated process.

Mr. Gutsinda explained to me how he and his neighbors acquired rights to land in Kiyovu cy'abakene by creating titles with lower-level city authorities. "For example," Mr. Gutsinda explained, "if I have land, you can tell me, you say 'I want this plot here . . . how much is it?' We talk and see if we like each other first because we are going to be neighbors, you understand?" They would be neighbors because, unlike land transfers in the city's cadastre, in which a seller transfers an entire plot to the buyer, landowners in Kiyovu cya'abakene and Kimicanga would divide their parcels. The seller would stay, and the buyer would move into the new plot and build their house. "Then," Mr. Gutsinda went on, "maybe we will share 'a bottle' and talk about the price." Then, as now, this process of transferring property over "a bottle," rather than through the impossible bureaucracy of land offices, happens under a name that can also be used to describe other deals made over a conversation: *inzoga z'abagabo* (beer for men).[5] Once the price and dimensions of the plot are agreed on, more people are called into the process—usually three more families, who are paid "beer" (money) to witness the sale and vet the new neighbors. The new landowners and their neighbors then create their own paper title to provide legal proof of their purchase. These titles record the dimensions and location of the new parcel and are signed by the three families who served as witnesses to the transaction.

Titles made through inzoga z'abagabo, while never recognized by the national government, can take on legal weight as low-level officials are brought into the process and induced to recognize them. By the 1970s, the government of Rwanda had installed cell (neighborhood-level) offices staffed by ruling-party secretaries, who were (as local leaders are today) charged with bringing the state into the popular city in Kiyovu cy'abakene and the neighboring neighborhood of Kimicanga. These officials were charged, in part, with enforcing the ruling party's prohibition of unauthorized growth, but they could be paid to collaborate with city residents to make their otherwise unofficial land titles official. Ms. Kampire, who moved to Kimicanga (a neighborhood that bordered Kiyovu) from Cyangugu, on the Rwanda-Congo border, in the 1970s, described the process of bringing intermediary state officials into land reform: "Then [in 1971], there were *abayobozi bari baciriritse* [mid/low-level leaders] who lived in the neighborhood . . . we paid them their *inzoga* [laughter], and they would sign our document to make it law" (recorded interview, March 8, 2015). The title would then be logged in the secretary's book, making the new parcel official (see figure 4).

As the road workers extended the airport road west and then back, crossing the wetlands, they wrote their own cadastre, building the city with clay and documenting their own land transactions. The popular city then pushed back along the northwest edge of Nyarugenge Hill, always between the tarmac road and the wetlands. Cyahafi came up and ran through Kandt's former residence and then spread south along the tarmac up through Gitega. As these neighborhoods surrounded the colonial core, they merged with Biryogo and Tapis Rouge, the former Swahili

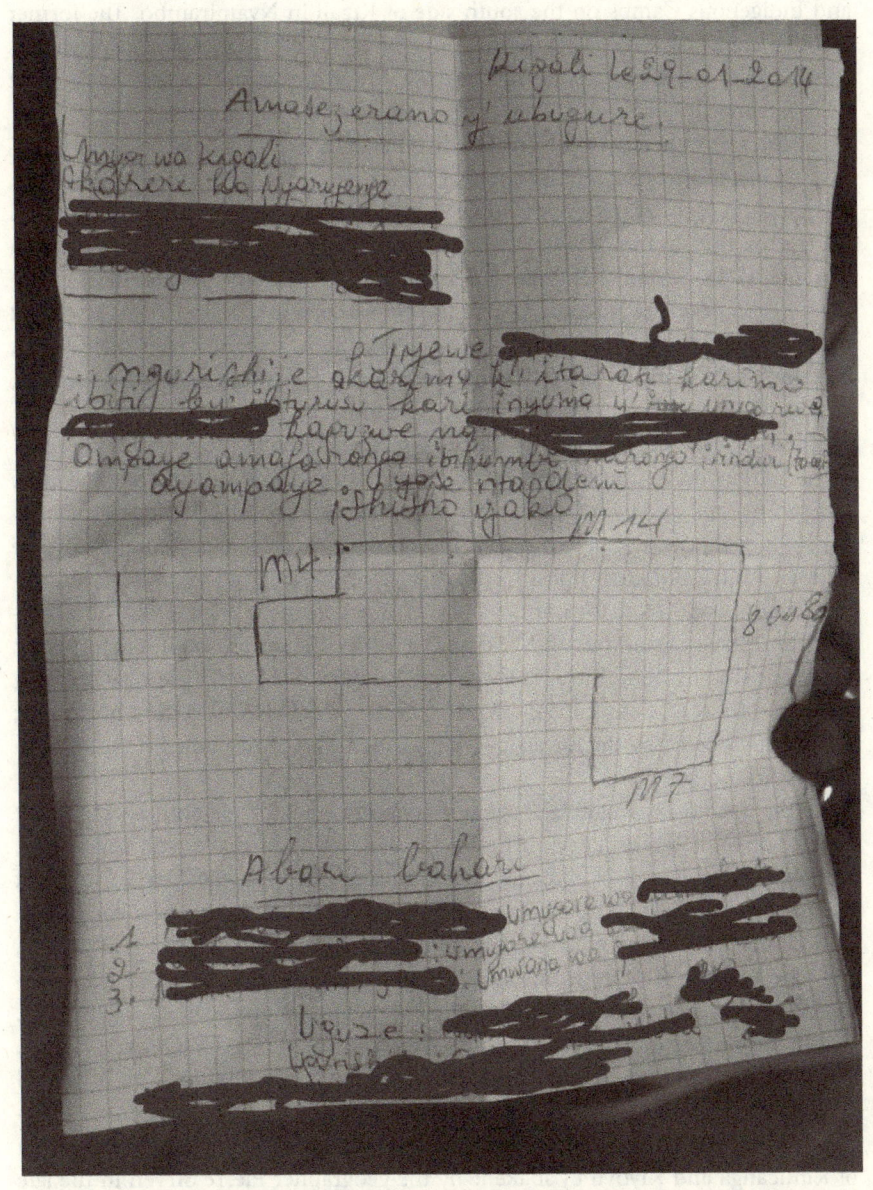

FIGURE 4. A more recent title made through inzoga z'abagabo. Names and addresses have been redacted for security. The title of the document is "amasezerano y'ubugure" (contract of purchase). Below the title are the precise locations (city of Kigali, District of Nyarugenge, followed by the sector, cell, and neighborhood names), and at the bottom are the signatures of the parties to the transaction and the three witnesses. This title contradicts official accounts that "informal" titles had no legal or monetary value. The land in this title was bought for RWF 70,000, as stated in the fourth line of the second paragraph. Photo by Samuel Shearer, 2014.

and Indigenous Camps on the south side of Kigali in Nyamirambo. The former colonial core of the city became surrounded by an African-owned city, with its own parallel system of titling, made up of single-story courtyard housing built using (primarily) materials—clay, sand, earth—from the wetlands.

By the 1970s, the Kayibanda regime had fallen into disarray due to elite infighting and lost most of its popular support (Desrosiers 2014, 206; Kimonyo 2008, 58; Lemarchand 1970, 238). In 1973 Minister of Defense Juvénal Habyarimana seized control of the Rwandan state in a coup d'état. Two years later, Habyarimana announced the Second Republic and a one-party development state, the Mouvement Révolutionnaire National pour le Développement (MRND). He was rewarded with an outpouring of development and military aid from France, the United States, and the World Bank. In the following years, agriculture modernization programs, funded by World Bank loans, gave rise to a booming coffee export industry.

The Habyarimana regime was, like its postindependence predecessor, populist-authoritarian, ethnocentric, and anti-urbanist. The political rhetoric of the regime was pro-peasant and packaged in the pseudo race science inherited from colonial-era science fictions. Habyarimana's discourse professed abandonment of the "feudal" (Tutsi monarchy) past in the service of modernization while maintaining "traditional" (Hutu peasant) values. The government tried (with little success) to restrict urban growth by requiring all Rwandans to obtain official proof of formal employment to reside there (see Vandersypen 1977). The new building codes and passes, as one consultancy observed in the 1980s, were simply ignored, mostly unenforced by the lower-level authorities tasked with doing it (see Rivkin Associates Inc. 1983, 9).

In the meantime, Kigali landowners worked to make their homes permanent sources of value by installing infrastructure. In a process that is still used today, landowners in Kiyovu cy'abakene and Kimicanga built on the long histories of using collectivization to exploit labor in Rwanda to make their own savings association. The association of property owners raised money to pay ELECTROGAZ, the state utility company, to install water and electricity in their new neighborhoods. As Mr. Nsengimana remembered, "we [landholders] worked together . . . we put our money together and then we went and asked ELECTROGAZ . . . more people joined us in the association, and we all made contributions until we had enough to get rid of our Colemans and go electric [laughter]" (recorded interview, February 8, 2015). Mr. Nsengimana's memory is backed up by photos taken of Kimicanga and Kiyovu cy'abakene by the geographer Pierre Sirven in the late 1970s and early 1980s that show electricity and drainage infrastructure already installed in these neighborhoods (see Sirven 1984, 547, 549).[6] As the popular city grew, Kigali's population surged from an estimated 54,000 in 1970 to 117,749 in 1978 (Olson 1990, 34).

It is worth pausing to look more closely at the historical significance and everyday politics of city building that gave shape to late twentieth-century Kigali and the heterogeneous authorities that governed it in relation to other work on African urbanism during the same time. In her work on shack dwellers in Cape Town from the 1970s to the early 2000s, Anne-Maria Makhulu shows how the everyday work of settlement and homemaking during and immediately after apartheid both transformed the shape of Cape Town and facilitated political struggles against apartheid in ways that could not be matched by the African National Congress. As Black Capetonians cleared land, formed associations, and built homes, they engaged in "a politics of presence" against the apartheid state's attempts to keep non-white South Africans out of the city—making freedom by making homes. Unlike the celebrated "ideological heroism" of the official liberation movement, this was a politics that "would take on a fundamental issue of the right to the city" through "ordinary strategies of homebuilding, affective labor, [and] community organizing . . . " (Makhulu 2015, 109). This argument is compelling and aligns with an extensive scholarship on how people negotiate urban belonging and assert their right to the city outside of formal civil society (see Chatterjee 2004; Gastrow 2017b; Holston 2008; Kinyanjui 2013). But unlike Cape Town, where city authorities engaged in well-financed and militarized strategies to block access to the city and where shack dwellers were without a doubt resisting those strategies, early-independence Kigali had no real influx controls to resist—just building codes and vague pass laws that were unenforceable because the very people who were tasked with policing the city (lower-level authorities) were participating in unauthorized construction. Twentieth-century Kigali also does not fit neatly into the situation that Filip de Boeck, in his 2011 work on Kinshasa, and Jacob Doherty, in his 2022 work on Kampala, both call "laissez-faire" urbanism. Apparently, in twentieth-century Kinshasa, the government was too dysfunctional, and in Kampala before Museveni, it was too distracted by civil war to regulate space, so people built wherever they wanted.

In Rwanda, however, the Kayibanda and Habyarimana regimes were involved in the lives of their citizens and flush with financing from external donors that they used to implement plans. Their attention was simply directed elsewhere, toward rural areas. And in Kigali there were already well-established architectures and land transaction processes that did involve state agents such as cell authorities who regulated who built what and where. What was missing were the reforms promised by decolonization: land reform that would allow Rwandans to own property and build in the city, the extension of services beyond the colonial core, and new housing for new arrivals.

In his work on the *suburbios* in early independence-era Maputo, Mozambique, David Morton describes a politics of city building that was similar to the ways Kiyovu cy'abakene and Kimicanga were built. Morton (2019, 17) describes how residents in early-independence Maputo "attempted to give substance to being new citizens of a new state . . . acting as if the government were intervening in their lives—executing housing policy and urbanizing neighborhoods—even as the attention of authorities was absorbed elsewhere."

Similarly, in late twentieth-century Kigali, landowners and lower-level state officials engaged with their built environment *as if* the national state was intervening in their lives and decolonization was a real project, not just a series of talking points. Just as Kandt organized the policy of shoring up the authority of the mwami *as if* he were the "natural" ruler of Rwanda who needed support from a German fort, just as the Belgians segregated central Kigali *as if* Islam posed a threat to their interests in the Congo, creating a city that was segregated on the basis of identities that were externally imposed on its residents, Kigali's early-independence workers proceeded to build the popular city as if the state had made efforts to decolonize the city, reifying the promises of decolonization to make it real.

Landowners like Mr. Gutsinda and Ms. Kampire did not fight the state but worked with lower-level authorities to innovate alternative titling schemes as if there had been land and property reforms at the national level that allowed Rwandans to enter Kigali and own land there. The new landowners hired abafundi to build affordable housing with clay from the ground below as if there were a national policy to extend working-class housing in the city using the most affordable (yet durable) building materials even though there was none. Landowners like Mr. Gutsinda took it upon themselves to form associations and collect money and then pay the utility company to install services as if there were a state program to collect surpluses from ground rents and use those surpluses to modernize the city by extending electricity and water infrastructure. In the process, Kigali's new residents, in collaboration with lower-level state officials, did achieve some of the promises of decolonization by making the fictive promises of land reform, affordable housing, and the right to urban space real. The result was not an ahistorical collection of "slums" or "informal settlements," but neighborhoods like Kiyovu cy'abakene and Kimicanga—places built through historically established processes of land transfers and clay architectures.

BOOM AND BUST

It was not until the 1980s that national leaders in the government of Rwanda became concerned about neighborhoods like Kiyovu cy'abakene. In 1981 the government of Rwanda turned its attention to housing as part of a twenty-year plan to accommodate new housing and commercial sectors and to stop popular neighborhoods from swallowing the former colonial core. To support the plan, the government of Rwanda issued decree no. 4/81, which, like the 1968 decree before it, made any structure built after October 1, 1967, without government permission illegal ex post facto (Sirven 1984, 531). But this plan "had almost no ability to influence economic activities, the settlement choices of inhabitants, or the directions of expansion . . . [in part because it was] hindered by a lack of adequate financial support and oversight" (K. Smith et al. 2018, 106). The law and the plan had little effect because there were no enforcement mechanisms in place to counter the processes of land transfer and clay building that Kigali residents had already established with local authorities outside of national laws and surveillance. The city's population surged again, to 235,664 in 1991, with almost all of this new outgrowth in the neighborhoods that surrounded the core (City of Kigali 2002, 11).

Some former residents of Kiyovu cy'abakene who lived in Kigali under the Habyarimana government expressed a certain nostalgia for the period—not because they were living under an ethnonationalist military dictatorship supported by Western donors, but because, thanks to the coffee boom and the outpouring of aid to the development state, the Rwandan franc was strong, there was work in the city, and taxes and the cost of living were low. "There was money [in the 1970s and 1980s]," Mr. Gutsinda, who identified as a genocide survivor and an ardent supporter of the Rwandan Patriotic Front (RPF), explained his nostalgia for this time to me. "You could go around lending money to friends, we had money to go out, to build my house . . . and they [city authorities] did not destroy [neighborhoods] a lot [*ntabwo basenye cyane*] . . . not like today" (recorded interview, April 17, 2015). As Mr. Nsengimana put it, "*mucaro* [in the country], we had to use Colemans [gas lamps], there was no electricity . . . but in Kiyovu we could go dancing . . . because I was a boxer, you know? There was Hotel Kiyovu, Kigali Nights . . . it was hot . . . at night, on weekends . . . like that." The popular city, for all its problems, offered access to an experience of electrified, independent modernity that was unavailable elsewhere.

But the Kigali nights wouldn't last.

. . .

In 1986 Cold War capitalism came crashing down on Kigali, and all that money lost its value. The global coffee price crashed and took the Rwandan franc with it, leaving the government of Rwanda with debt in dollars that the franc could no longer repay. Popular support for the Habyarimana regime began to wane, with regional factions springing up throughout Rwanda and the opposition of exiles in the Rwandan diaspora growing. In 1990 the Rwandan Patriotic Army (RPA), the military wing of the opposition in exile, invaded from Uganda, led by a cadre of mostly high-ranking Rwandan soldiers who had been trained in Yoweri Museveni's military and police forces.

By that time, national income from coffee exports had dropped from $144 million at its height to just $30 million (Mamdani 2001, 149; Uvin 1998, 11). The Rwandan franc lost 67 percent of its value. The price of imports skyrocketed, increasing sevenfold, while the national debt ballooned. Famine hit rural Rwanda, and unemployment soared in the city. As the civil war unfolded, the military recruited unemployed young people into Interahamwe (youth militias), ostensibly to "protect" Rwanda from the RPA but in reality to do the work of protecting ruling elites from the fictitious threat of the Tutsi, who had been scapegoated as an internal enemy. By April 6, 1994, when a plane carrying Juvénal Habyarimana and his Burundian counterpart was shot down over Kigali by an unknown agent, the conditions for genocide were in place.[7]

Over the next three months, the 1994 genocide against the Tutsi unfolded across Rwanda. The military and Interahamwe murdered around eight hundred thousand people. As the front lines in the civil war moved into the city, many Kigali residents walked away from the murder in the streets and the shells dropping from the sky to camps in Zaire (present-day Congo), Tanzania, and Burundi. The city's population shrank. On July 4, 1994, when the RPA took Kigali, effectively ending the genocide, and merged with its political party, the RPF, only an estimated thirty thousand to ninety thousand people remained in the city (de Montclos 2000, 12).

Rose Kabuye, a lieutenant colonel in the RPF, was installed as mayor by the interim government. Her priority was crisis management. Corpses had to be removed from view, identified, and buried. Water and electricity had to be turned on and the more immediate tasks of peace and reconciliation had to be attended to. In the meantime, the RPF had its hands full fighting insurgency in the north and supporting Congolese rebels as they toppled Mobutu Sese Seko in Zaire. Kigali was essentially a city under military occupation run by Rwandans who had spent most of their lives as refugees in Uganda.

Kigali authorities also had to deal with surging population growth. As the war ended, former Kigali residents returned from refugee camps in Zaire and Burundi. "Old caseload" refugees—Rwandans who had fled the Kayibanda and Habyarimana regimes in the 1950s, 1960s, and 1970s—also returned from Uganda and Tanzania. From 1995 to 2010, Kigali would become the fastest-growing city in the fastest-urbanizing country in the world (see K. Smith et al. 2018, 49).

At the same time, thanks to a crashed rural economy and an intermittent war on the Rwanda-Zaire border, thousands of new arrivals from rural areas also came to the city. In just six years, the population of Kigali grew from 30,000–90,000 in 1994 to an unprecedented 604,966 in 2000 (KIST 2001, 3). The population continued to grow, to 800,000 in 2004, a decade after the genocide (Manirakiza et al. 2019, 291). During this time, ruling party leaders simply did not have the resources to handle even routine day-to-day tasks such as tax collection, building permits, and business registration (see KIST 2001).

Consequently, the city's hundreds of thousands of returnees and new residents were tasked with repairing their own city.

POSTGENOCIDE REPAIR

Postgenocide Kigali was damaged, and it had its share of issues with housing, infrastructure, and security. But the city was not a blank slate. In 2015, when I first met Ms. Kiyobe, she was the head of a four-generation household located in Kiyovu cy'abakene. We were connected by her former neighbor, Ms. Ingabire (see chapter 3), but I had been fascinated by her house for much longer. I walked by it almost every day on my way to Nyabugogo Market. It sat on the edge of a field where Kiyovu cy'abakene had once stood, spared from demolition in 2008. On one side was a vacant field; on the other side was Ms. Kiyobe's house, connected to a dense popular neighborhood of large, medium, and small houses, shops, and bars that stretched for several kilometers to Nyabugogo.

Ms. Kiyobe was born in eastern Congo and moved to Kiyovu cy'abakene in 1972, where she married a Rwandan construction worker. She worked in the neighborhood market that was located downhill from their house. She and her husband had started with a single room wattle-and-daub structure with a stamped-earth foundation. As they lived their lives, and invested money into their house, they removed walls, made extensions to the structure, replaced the stamped-earth foundation with stone and concrete. In the late 1980s, they put up an iron gate around the property and built two rental properties with amatafari ya rukarakara.

Ms. Kiyobe lost her husband in the genocide. When the fighting came to Kigali, she fled to Congo. She returned to Kiyovu cy'abakene in 1995.

"My house was still here," she explained. She added, with irony, "We lost family, but my house was in good health [*twasanze inzu yacu ari nzima ariko umuryango twarawubuze*]" (recorded interview, May 21, 2015). It had been ransacked, and many of her neighbors' houses had been demolished, but she still had the documentation to prove it was her house. She moved back in.

Ms. Kayirebwa had a similar story. She was born in Muhima, a neighborhood bordering Kiyovu cy'abakene, and had experienced its transition from *mucaro* (the country) to *mumujyi* (the city) with infrastructural connections in the late twentieth century. She and her five children fled the advancing RPA in July 1994.[8] After a two-week journey on foot to Zaire, they took one look at the camps—infiltrated with Interahamwe and festering with disease and famine—and walked back to Kigali. "When we left [Kigali]," she said, "the others came—refugees from [19]59. They came from Uganda, Burundi, everywhere." She added, "They had cash!" (recorded interview, April 30, 2014). And Ms. Kayirebwa needed cash.

When Ms. Kayirebwa returned to Muhima with her children, she divided her own plot to sell several small portions to "old caseload" refugees from Uganda, who built new housing. These land transactions were, as such transactions were in the past, recorded with nearby neighbors as witnesses. Ms. Kayirebwa then took advantage of the new market for affordable housing, building two one-room rentals on her plot, also with amatafari ya rukarakara, helping to resolve the postgenocide housing shortage, quite literally brick by brick.

Ms. Kayirebwa was not the only Muhima landowner who converted the postgenocide population boom into a real estate boom. Soon, Muhima, Kiyovu cy'abakene, and Kimicanga were packed with affordable rental properties built with materials from the nearby wetlands, owned by on-site Rwandan landlords. Mr. Nsengimana described the building boom of the late 1990s in Kimicanga this way. "In that time, . . . you look out your window and see your neighbor's house.

You go to sleep and then you wake up, and you don't know where you are because someone built another house two feet away while you were sleeping! [laughter]" (recorded interview, February 8, 2015).

These accounts of Kigali residents rebuilding their own city after the genocide are backed up by many others in the city, and by secondary sources. During the postgenocide period, the Rwandan geographer Vincent Manirakiza (2011, 12) describes: "Many people ignored urban regulations and built their own houses almost everywhere in the city. It was a remarkable period, in which existing neighborhoods were densified and new working-class neighborhoods were built below elite suburbs, on the slopes of Kimisagara and Gatsatsa" (see also Manirakiza and Ansoms 2014). The accessibility of clay, quick construction techniques, and parallel titling processes that built and sustained Kiyovu cy'abakene in the postindependence period provided the base that made it possible for Ms. Kiyobe and Ms. Kayirebwa to return to their properties and rebuild their lives in the city.[9]

Life during this time was, of course, rough. Ongoing regional violence and poverty were certainly drivers in Kigali's rapid growth. There were ample land and property disputes, not to mention issues with security, public health, public trust, and postgenocide trauma. The city's population was growing at an unprecedented rate, but its economy was not. It is therefore understandable why some policy scholars reproduce official accounts that "informal settlements" and "slums" haphazardly appeared during this time in a mode of "random settlement" (see Goodfellow 2022, 89). Indeed, there was more than one UN report that worried about the city's swelling numbers of "shantytowns" (see de Montclos 2000). But these reports miss that the city's neighborhoods were being built on preestablished knowledge of how to work with the city's ecologies to put a built environment together in ways that also put urban authority together. That knowledge followed systematized construction and contracting practices: the mechanisms of clay construction and inzoga z'abagabo, established in the early independence-era city.

As landowners like Ms. Kiyobe worked on resolving the postgenocide housing shortage, mechanics, tailors, and other craftspeople also worked with what was already there. Tailors and cobblers took the wasted clothing being dumped in Kigali's streets from rich-world economies and remade it into new fashions. Produce vendors took vegetables on the verge of becoming waste and sold them in small units at a discount. The result of all this activity was a growing street economy of discounting and recycling that spilled back from the wetlands and into the city center, taking over the former colonial core.

Mr. Imanzi, a secondhand clothing trader who let me shadow his work in Nyabugogo Market, recalled the atmosphere in Kigali when he arrived from a small town in the south in the late 1990s: "It was hot [*harashyushe cyane*]! Listen [*umva*]: I had my box, I put it on a crate every day. It was 'Manzi's Place' [*kwa Manzi*] [laughter]. I sold peanuts, cigarettes, tissues, gum, cookies, chocolate, like that . . . ." (recorded interview, July 6, 2014). Mr. Imanzi was twelve years old when the RPF marched on Kigali in July 1994. He managed to survive the genocide with his mother by hiding from the militias in the forests outside his hometown. After the genocide, like so many of his generation, he headed to the city. He needed a job: "It was like you could build anything, anywhere then. You could set up a small shelter on the side of the road and then it was your kiosk, okay? [He used the French *bon*.] Sometimes, soldiers would come by and tear down a shelter, or sometimes you would have to pay a little something [*akantu*, a small bribe, literally "a little something"], but mostly, it was just business [*ubucuruzi*]." Mr. Ishimwe, a thirty-four-year-old *umuzunguzayi* (mobile street trader) when I spoke to him in 2015, agreed: "*Sha*, in those days [the late 1990s], it was just one big market in the city: the bus station, Commercial . . . Nyarugenge . . . You had every type of merchant selling everything. We were walking in the streets, behind us, you had little shops selling clothes and electronics. There weren't buildings like now, just people" (recorded interview, March 30, 2015). These stories are about survival in precarious times. But they are also—in a very different way—about what Shakirah Hudani (2024) calls "the material politics of repair" in postgenocide Rwanda.

To repair their lives and make the city work for them, landowners like Ms. Kiyobe and Ms. Kayirebwa and entrepreneurs like Mr. Imanzi and Mr. Ishimwe and the tens of thousands of other residents who poured into Kigali had to work with was already there—clay and earth, trashed automobiles, secondhand clothing, kilos of food about to rot. They then stripped the cloth, the automobiles, the kilos of tomatoes down to component parts while repurposing them to other things: resewn fashions, minibus taxis, and new housing stock.

MILLENNIAL DYSFUNCTION

The year 2000 marked a turning point for the ruling RPF, and for Kigali. The government of Rwanda released its lauded Vision 2020, a national development plan that aligned the government's priorities with the United Nations Millennium Development Goals. General Paul Kagame, the head of the Rwanda Defence Force and the country's de facto leader, became the official president of the Republic of Rwanda, further consolidating the ruling party's hold on national power. The RPF committed to a program of "decentralization," which meant building on previous regimes' networks of distributed surveillance that, in theory, run from the top—ruling-party headquarters—down to the umudugudu (village or community) level (see Purdeková 2011). As the RPF grew in political power and established security, government elites turned their attention to Kigali as a source of capital (see chapter 1).

As Tom Goodfellow writes, this was when the "RPF turned its military intelligence levers towards enhancing urban security as a central part of its legitimation drive, further deepening the state's existing infrastructural power through an intensification of community-based policing" (Goodfellow 2022, 89; see also

Lamarque 2020). According to Goodfellow, the RPF took control of the distributed network of local governing institutions and its diverse range of community police units set up to patrol urban space. But this urban control, however crucial to establishing security, was also more an aspiration than a reality.

For example, much has been written about the RPF's heavy-handed efforts to take control of Kigali streets in the early 2000s (see Hudani 2024, 76; Sommers 2012, 170). Police and security forces began raiding the central business district. They rounded up street children, street traders, and suspected sex workers. But many people who were in the way of the RPF's early attempts at controlling the city experienced this moment not as a period of government control, but as random municipal violence that was sometimes laughably ineffective.

According to Mr. Ishimwe, the RPF campaign to clear the city center began with the mayor making radio announcements urging street traders to leave the central business district. When that did not work, the Rwanda National Police, assisted by community policing units (sector-level security guards, then called local defense forces), began regular raids on the central business district.[10] The raids, however, immediately created new problems for the city: what were police and security guards supposed to do with street traders when they caught them? There was no room in Kigali's prisons, which were overcrowded with *genocidaires*, so the police improvised a detention facility: the Containers.

The Containers were actual shipping containers used during the early days after the genocide to hold the overflow of suspected *genocidaires*.[11] Beginning in 2001, the Rwanda National Police started using the facility to warehouse residents picked up in street raids, mostly for the crimes of selling food and recycled clothing in the streets. As Mr. Ishimwe remembers:

> *Sha*, if you came from the Containers, everyone could tell where you had been. They had these bedbugs, like this [laughing, holding his thumb and forefinger half an inch apart]! I am telling you, God's truth [*n'ukuri kw'imana*], that big! So if you wanted to lie—say your mother wanted to know why you hadn't come to visit her, and you said, "I had something to take care of out of town," she would know you were lying because you were *covered* in bedbug bites. [Shaking his head. We are both laughing.] I am not joking though! They would just throw everyone in and lock the doors. . . . When you got out . . . you would be such a mess, the wind could knock you down." (recorded interview, March 30, 2015)

These early raids on Nyarugenge produced a lot of bad memories, but like previous efforts to contain the city during colonialism and the early independence era, they were also ineffective. Rwandans who worked the street economy simply learned to run from community police units, calculating arrest into the cost of doing business.

And business was good. Minibus operators continued to drop customers from all over the region into the former colonial core in Nyarugenge, while adjacent neighborhoods like Kiyovu cy'abakene and Kimicanga kept the popular city thriving with a large stock of affordable housing. And at this writing, efforts to remove street economies from where Nyarugenge Market once stood continue to hopelessly fail.

These efforts failed not because of the heroics of "informality" or the inadequacy of policing, but because street economies of food, clothing, and housing are essential to the maintenance of the entire city and the authorities that govern it. As I show in chapter 5, the street economy is a discount economy that is essential to maintaining the vital but low-paid workers in every department from sanitation to education that keep Kigali running (see also Shearer 2020a). Similarly, the house built on Mount Kigali in 2013, like all the houses in this chapter, may have been out of code, but the house was much more than a shelter: it materialized the very institutions that disobeyed one policy (zoning) to support another ruling party project (the provision of housing). The enforcement of zoning failed so the provision of services could succeed. Nothing was improvised or random. It was built on a hundred-year history of putting heterogeneous authority together by putting the city together in ways that continue to confound efforts, no matter how authoritarian or violent, to shift the city in new directions.

These were the actual conditions of Kigali when the RPF leadership hired OZ, the first firm to redesign Kigali's future as a finance-driven, growth-based "sustainable" solution to the climate crisis: a city that had already been rebuilt after the genocide by the people who lived there through a continuum of long-standing knowledge of construction and repair. OZ's clients were ruling party executives who had only nominal control over the heterogeneous institutions of urban authority that were meant to police and control the city. At the same time, the strategy to sell Kigali to outside investors depended on the city being conceived of by outside investors as remaining in postgenocide hypostasis over a decade after Kigali residents had already rebuilt their own city. To make this fiction real, the repair work completed by landowners like Ms. Kiyobe and Ms. Kayirebwa through historically established land transfer and architectural processes would need to be undone, and the city would need to be destroyed all over again.

CONCLUSION

This chapter has tracked the history of Kigali's segregated built environment, racializing fictions, and the popular processes of construction and repair that not only make, maintain, and repair the city but also make and support the authorities that govern it.

Unlike colonial and independence-era governments, the RPF would direct considerable attention toward developing Kigali into a center of global finance and tourism. Also, unlike the previous plans that were based on twentieth-century ideas about the "rational" organization and governance of urban space, the OZ plan would be directed toward reaching an external audience that wanted to achieve a new urban quality—a quality called sustainability—so that the city could become a source of finance. This required more than planning. It required reframing the popular city—made mostly from low-carbon, locally sourced materials in dense, walkable neighborhoods and based on discount economies—as "unsustainable" and therefore in need of being replaced. It would require new institutions and laws that would surface fundamental contradictions at the core of market-driven sustainable urbanism, rendering upper-level officials surprisingly incapable of enforcing building and zoning codes. And it would inspire new travelogues written by visitors to Kigali. This time, these tales would be about a city rising from the embers of catastrophe into a green future.

3

Brand

Ruled by Fictions

On March 8, 2020, *National Geographic Travel* ran an article titled "Kigali: How Creativity Has Transformed the Rwandan Capital," by English travel writer Emma Gregg. The article's teaser reminds readers that while they might be planning a trip to Rwanda to see the famous gorillas, "the progressive capital has an allure of its own" because "the city is finally emerging from the embers of its past" (Gregg 2020).

Gregg's tour of Kigali focuses on the city's meeting, incentive, conference, and exhibition (MICE) district, a twelve-kilometer string of high-end hotels and conference centers that runs from the airport to the city's central business district. She begins her tour at Heaven, an exclusive boutique hotel/restaurant located in the upscale neighborhood of Kiyovu, the former European Camp that retains much of its original character. Heaven is owned by two US entrepreneurs who have relocated to Rwanda. "It's one of those restaurants-with-rooms," Gregg observes, "that attract movers and shakers, from visiting doctors, engineers and renewable energy experts to local crop scientists." When Gregg visits, many of Heaven's guests are there to attend one of several conferences being hosted that week in the freshly painted MICE district. "Rumour has it," Gregg writes, "that, come the Commonwealth Heads of Government Meeting 2020 in June, Prince Charles will be checking in [to Heaven]." The most high-profile of these events are held at the Kigali Convention Center, which is "domed like a traditional Rwandan palace but lit up like a sci-fi space station." Construction on the convention center was finished in 2016, at a cost of $300 million in public debt.

Kigali's allure, Gregg emphasizes, is the product of the city's postgenocide reinvention. Gregg finds evidence of a city being rebuilt from the embers of the past everywhere she goes—for instance, in Kigali's Car Free Day. Since 2016 Kigali authorities have closed a main thoroughfare to vehicle traffic one Sunday a month to encourage expatriates, tourists, and residents to exercise in the streets. The initiative, Gregg informs her readers, has been "[key] to the city's healing process" (2020). It is all part of a "new city" with "green credentials," "one of Africa's most livable." What is absent in Gregg's reporting of sci-fi space stations rising from the embers of the past into a green future is any sense of lingering trauma. To the contrary, quoting ample sources, from foreign business owners to Rwandan artists, Gregg conveys a sense of contagious utopian energy bubbling from the ruins of catastrophe in the Rwandan capital. And while every city in the region has a plan to become "clean" and "green," Gregg and her interlocutors confirm that Kigali is unique and different. Kigali is special, a Rwandan creative tells Gregg, because unlike "other African cities," Kigali is "not chaotic—in fact it feels really safe" (Gregg 2020).

Gregg's enthusiasm for Kigali's allure is not singular. Her work echoes what global development gurus like Tony Blair and Jeffrey Sachs call a "remarkable recovery" in Kigali (see Allison 2017). International organizations agree. UN publications describe how Kigali has transformed from a postcatastrophe city to a market leader in investment-driven sustainable urbanism and high-end conference tourism (Bafana 2016; Twahirwa 2018). Kigali's "clean and green" postcatastrophe recovery has taken the city's name to the top of the World Bank's "Ease of Doing Business" rankings (World Bank Group 2020), UN Millennium and Sustainable Development Goals rankings (Bhowmick 2019), and—most importantly—international bond and credit ratings (Wallace and Pronina 2014).

From a certain perspective, this narrative is correct. Much of Kigali's core was reduced to embers. But these demolitions began in July 2008, not 1994. And they started across the road from where Gregg stayed at Heaven, in Kiyovu cy'abakene, where hundreds of families were evicted from homes and businesses that were then demolished (see figure 5). Two years later, in 2010, more than a thousand homes and small businesses were demolished in Gacuriro, a neighborhood next to the convention center between Kiyovu and the airport (see figure 6). The destruction of two markets and more than two thousand homes and shops in Kimicanga, a neighborhood across the street from the affluent expatriate neighborhood of Kimihurura, followed in 2012 (see figure 7). Lafreshaîre (a neighborhood moniker) and Nyabugogo Market were demolished in 2014 (see chapter 6), followed by Kazaire in 2020 and Bannyahe in 2022 (see chapter 5).

What Gregg witnesses and reports back about is real, but not in the fundamental sense of the term. It is the materialization of a brand image that is sold through the city's MICE district and based on what external audiences expect to find on the ground: a postcrisis tabula rasa that is also a competitive opportunity to experiment with the best practices of finance-driven sustainable urbanism. In 2005 that brand was commissioned by the government of Rwanda, which hired a US firm to write the first sustainable plan for the city and involved foreign firms like OZ, UN reports, and globally circulating ideas about how to achieve a quality called

FIGURE 5. Kiyovu cy'abakene before (*top*) and after (*bottom*) it was demolished in 2008. Source: Google Earth, Digital Globe.

"sustainability," in which the climate emergency is conceived as raw materials for new markets.

Chapter 2 tracked the history of popular city building in Kigali: the land transfers, architectures, and economies that built neighborhoods like Kiyovu cy'abakene that were then used to repair the city after the 1994 genocide against the Tutsi. This chapter is about the Kigali City Master Plan, how it came into being, and the conflicting ideas and interests that continue to drive green capitalism in Kigali.

FIGURE 6. Gacuriro before it was demolished (*top*) and after the 2010 demolition (*bottom*). Source: Google Earth, Digital Globe.

I argue that the destruction of neighborhoods like Kiyovu cy'abakene (and dozens more) after they were already rebuilt is more than the creative destruction of urban renewal. The destruction of these neighborhoods is essential to the design, manufacture, and maintenance of a ruined cityscape, which is presented in planning documents and promotional materials as an opportunity to build a better, more ideal green city that can withstand the shocks of climate change. That brand image is the centerpiece of planning in Kigali, which is, in the words of the official

FIGURE 7. Kimicanga before it was demolished (*top*) and after the 2012 demolition (*bottom*). Source: Google Earth, Digital Globe.

documents that activated a search for an outside firm, to "[m]arket a stronger image of Rwanda/Kigali as an exciting, safe, affordable visitor destination, and in particular market Kigali more effectively to potential regional visitors" (City of Kigali 2002, 79; see also Goodfellow and Smith 2013).

Urban plans, of course, are much more than brand images and marketing materials. Despite being delivered by experts and taking on the appearance of rational designs, plans rarely take shape as coherent projects. Rather, they are given form through "debates and dissension, struggles over bureaucratic power" (Bissell 2011,

18). As I show in chapters 4, 5, and 6, Kigali's plans are also used as instruments that go beyond the digital image. They are wielded by city officials, international hedge funds, and foreign consultants to enact their own agendas. They take on the appearance of tabulae rasae but are assembled from found parts: previous reports on housing, other plans that never materialized, and past dreams. They give way to new institutions and require tools for implementation such as statistical projections, ancillary reports to support accounts of a city and its future (Holston 1989). In Kigali, what began as a project to put together a single document of less than two hundred pages has become a sprawl of subarea master plans, comprehensive plans, land tenure regimes, expropriation laws, district plans, digital zoning databases, and housing market reports generated by institutions and consultants that were often competing with each other as much as they were trying to give form to the city.

While the rest of this book takes place in neighborhoods, streets, and markets, where planning is implemented and contested, this chapter aims to surface the entangled genealogies of planning as marketing, as crisis-driven market creation, and the debates and dissention that made the first sustainable plan for Kigali. To take Gregg's sci-fi analogy a step further, plans are texts that emerge from specific contexts. They are written by people who come from far-flung corners of the world and who have their own visions of what the ideal city should look like. They include projections of the future (the scientific reporting of fictions as facts), they have plot points and narrative arcs, and they contain ideas about the environment and its relationship to the city and the people who live there. Surfacing those elements helps shed light on the specific style of planning that the government of Rwanda brought to Kigali at the turn of the twenty-first century while offering a framework to understand the heterogeneous operations that go into struggling over and envisioning Kigali's future.

THE GREAT SELLOUT

As discussed in chapter 1, when President Kagame met Carl Worthington in 2003, the ruling party of Rwanda was looking for a firm that could market Kigali as a competitive destination for foreign investment. Kagame is on record in interviews and speeches that span several decades criticizing aid as a soft-power tool that undermines national sovereignty. He often cites private foreign investment as a preferred development strategy that, in theory, delinks state finances and policies from the soft-power strings attached to aid (see Moyo 2009, 27; J. Fox 2013).

During their initial meeting, Worthington took Kagame and his cabinet ministers to his office overlooking the Denver Tech Center, a private technology park Worthington designed in the late 1960s south of Denver's then-crumbling postindustrial downtown. As Worthington explained to a US reporter after the plan went public, he offered Kagame the investment-driven alternative to aid. This was the alternative that the president wanted: "At the end of the slideshow I said [to Kagame and his cabinet], 'We are now looking down on 20 million square feet of development that has been all privately done. It started with a vision, then the money came, then the corporations came.' That is what sold the president. He turned to me and said 'Carl, would you like to come to my country and help me rebuild it?'" (Marshall 2008, 26).

Because Kigali has a history of conflict and because the ruling party of Rwanda provides the lens through which much of the world understands Kigali, the Kigali City Master Plan is often read through a perspective of postconflict nation and state building with emphasis on James Scott's (1998) theory of twentieth-century modernist projects (see Hudani 2024; Goodfellow 2022). For example, the historian Timothy Longman claims the OZ plan for Kigali is a "fantastical example of Scott's high modernist vision" (2017, 178; see also Sommers 2012, 207). This is because, according to Longman, the plan "involves a radical reconceptualization of Kigali as a carefully planned city along the lines of Brasilia, Dubai, or Abu Dhabi and includes drawings of gleaming glass towers and pristine shopping districts along tree-lined boulevards" (2017, 178–79). But the Kigali Conceptual Master Plan has none of those features (although later YouTube marketing videos do). Kigali, the city, and its plans are nothing like Brasilia, a federal capital built out of concrete in 1957 long before sustainability entered the vocabulary of architects or planners (see Holston 1989). Nor is the Kigali Conceptual Master Plan anything like Dubai, a theme park of spectacles designed by celebrity "starchitects" in the late twentieth century and paid for with surplus rents from oil (a resource that Rwanda does not have) (see Kanna 2011). Unlike the aspirations to remake cities as holistic and legible that were a feature of high modernism, the Kigali Conceptual Master Plan is a series of divergent programs, implemented ad hoc and driven by debates about how to attract investment to the Rwandan capital. In addition, as other scholars of planning in Africa have observed, Scott's formula of high modernist control "grants too much power and rationality to the state ... accepting it as reality rather than representation" (Bissell 2011, 113).

Furthermore, plans are written to persuade. And as every architect today knows, the cold-war high modernist city that features in James Scott's work has been "defunct for several decades now ... Perhaps it is still called upon as a repertoire of pure form, its ideological basis just a parenthetical warning to true believers ..." (Chattopadhyay 2012, v; see also Buck-Morss 1995, 2000). References to Kigali's future as a copy of Singapore (or Dubai, or Shanghai, or more recently the fictional city-state of Wakanda) are, as Asher Ghertner observes in similar strategies to align Delhi with Paris, simply "diffuse signifiers" in which "any other iconic city might just as easily stand in" (2015, 9).

What we do learn from the serious scholarship on twentieth-century modernist city building in Africa is that modernism in Africa "emerged in parallel with, not as a 'secondhand' latecomer copy of the metropolitan model with which it was co-incident" (Chalfin 2023, 45; see also Bissell 2011; Elleh 2002; Hoffman 2017; Rabinow 1989; Wright 1991). Similarly, the Kigali Conceptual Master Plan, a twenty-first-century sustainable plan, emerged contemporaneously with other sustainable plans around the world in the first decade of this century (see Carmondy et al. 2023; Murray 2015; Myers 2016 on sustainability in other African cities).

Unlike the modernist project—in which building a new city was supposed to create a new society—the Kigali City Master Plan belongs to a style of urban planning and management that aims to convert the crises of economy and ecology into new growth-based markets (see Greenberg 2015 on New Orleans; Checker 2020 on New York City; and Günel 2019 on Masdar City).

One of the most insightful accounts of twenty-first-century market-driven sustainable urbanism in Africa, and the predicaments that drive it, comes not from studies of modernist planning (as important as those studies are) but from the Zimbabwean novelist and sci-fi writer Tendai Huchu. In his short postapocalypse story "The Sale," Huchu describes a future African metropolis in which all the promises of green city building have been achieved. In Huchu's future, our present (the first decades of the twenty-first century) has come to be known as the "great sell-out." In those decades, "Third World nations heavily under debt were sold off piecemeal to Corporations [sic]" (Huchu 2012, 34). This led to a consolidation of urban governance in which future cities are run by a global public-private partnership aptly named CorpGov: a hybrid technocratic state and investment corporation that bears an uncanny resemblance to the ruling party of Rwanda.

As "The Sale" unfolds, Huchu walks his readers through a future capital city run by CorpGov and its subsidiaries. CorpGov exercises total control over urban space, as well as the labor, movement, and birth rate of Harare's "native" population. In "The Sale," everything that CorpGov claims to be able to fix gets fixed through a "stable equilibrium combining free markets, scientific rationale, and central planning" (Huchu 2012, 40). The city's inhabitants have twenty-four-hour water and electricity service, provided by private corporations; universal health care is administered by drones (who also control the population and sterilize "natives"); life expectancy is up by 100 percent. The one issue that remains—despite the city's green infrastructures, including electric rail and hydrogen-powered automobiles—is the "thin brown toxic haze" that obscures the sun every morning (2012, 33).

The world that Huchu builds in "The Sale" could also be read as an example of high modernist ambition: a space of total control governed by ideas of scientific rationality. But the institutions and policies in "The Sale" are not directed toward making the unmanageable more manageable. They are about addressing a very real problem—debt loads inherited from the past—by selling the city out to corporations.

Like most great dystopian fiction, "The Sale" offers a critique of the present from the perspective of an estranged future. African cities like Kigali are struggling under heavy debt loads thanks to twentieth-century Cold War capitalism and structural adjustment programs, which were exacerbated in Kigali by the 1994 genocide against the Tutsi (see chapter 2). And, thanks to the emergence of foreign direct investment as a mode of development, many cities in Africa, like Kigali, are being run by municipalities that are using plans for sustainable futures to market their city to outside investors through public-private partnerships (CorpGov). But what makes "The Sale" dystopian—what makes it distinct from a standard critique of new plans for African cities as "fantasies" that "fail"—is that Huchu accepts the utopian premise of privatized sustainable city building. Huchu then builds a world in which a plan for a green future succeeds, identifying who will pay the costs of following that project through.

If Huchu were to tell the story of the great sellout from the perspective of *National Geographic Travelogue*, the narrative would match the golly-gee-whiz-isn't-this-amazing account in Gregg's reporting. The future city, from the perspective of the population that Huchu identifies as "Businessmen"—outside investors, tourists, a managerial class—is one of the most livable cities in the world. But "The Sale" is not told from the perspective of a visitor or a corporate consultant; it is told from the perspective of Mr. Munyuki, "a native." What Mr. Munyuki's perspective reveals is the price paid by the city's inhabitants for sustaining a "stable equilibrium" between zero-carbon capitalism and market growth: a city that is run by corporations rather than residents, where the people who build and maintain the city are written out of the future, and where the issue that green capitalism cannot address—the unequal distribution of a thin toxic haze—continues unabated. Furthermore, as Huchu shows in "The Sale," defining for *whom* the plan is utopian—external audiences of investors—makes the flip side of utopia available: the people who pay the material costs for the marketing fictions.

As a critical dystopia written to make a clear point, Huchu's work is Manichaean, presenting a world divided into "businessmen" and "natives" (see Moonsamy 2016). Of course, the actual processes of sustainable city planning are much more complex. There were no bad actors in making the Kigali Conceptual Master Plan, only consultants selling their labor to the highest bidder while trying to find solutions to problems that did not cohere.

If the Kigali City Master Plan were only a marketing tool, none of this would matter much. But as Huchu points out in his rendering of the future, reaping the benefits promised by finance-driven sustainable urbanism requires making the real city match the unique brand image by creating the postcrisis, tabula rasa Kigali on which the plan builds its pitch. This necessity leads to new contradictions. Aligning the city's built environment with the fictions external investors want to believe so the ruling party can achieve its goal of attracting outside financing and break free from aid (and the soft-power strings attached to it) requires demolishing actual neighborhoods, creating intractable challenges for city managers who still need to service the city and try to govern its form.

The next section then starts at the source of a specific approach to sustainable urbanism, which can be charted through the professional biography of the plan's lead architect, Carl Worthington. In following Worthington's career as it traveled from Boulder to Asia, then South Africa, and finally to Kigali, my aim is not to twin Kigali to Boulder (as though Kigali has been modeled on Boulder, not on Singapore or Dubai). Nor is it to criticize Worthington for doing a job he was hired by the government of Rwanda to do or to suggest he played a direct role in the demolitions that began once his job was finished. The aim is to trace the genealogical origins of market-driven sustainable urbanism from urban branding and crisis-driven urbanism to new strategies of making markets for green commodities.

GREENBELTS AND CATASTROPHES

Carl Worthington grew up in Los Angeles, the son of a successful real estate developer. He trained as an architect at the University of Southern California, where he received his bachelor's degree in 1961. His award-winning senior thesis was ahead of its time: a solar-powered luxury hotel designed to be carved into the side of a mountain—a high-end place for a transient population of visitors, run on renewable energy. After college, Worthington moved to his adopted hometown of Boulder, Colorado.

Boulder was founded in the late nineteenth century as a US mining settlement.[1] Between World War I and World War II, Boulder was not the haven of progressive politics it is known for today. It was openly run by a chapter of the Ku Klux Klan, which had its own press, "security" patrols, and seats on the city council (Hickcox 2017). Then, in 1952, ten years before Worthington arrived, Boulder was connected to Denver via a highway (the Boulder Turnpike) and the Rocky Flats nuclear plant opened nearby, bringing industry, research, and a new, diverse workforce to the area. In two decades, Boulder's population more than trebled, from 19,999 people in 1950 to 66,870 in 1970 (Pettem 2010).

In 1966, at the height of this population boom, Worthington joined Boulder's planning board. The board, along with a coalition of property owners, was nervous about the effects unrestricted development would have on the city's natural ecologies and its changing demographics. By his own account, Worthington played a key role in creating Boulder's celebrated 1967 greenbelt buffer zone (City of Boulder 2013). The planning board pushed through a one-cent sales tax increase. Boulder became the first city in the United States to use sales tax revenue to purchase fields, riverbeds, and wetlands that surrounded the city, creating a natural buffer that enclosed Boulder to prevent development from extending outward. Boulder's greenbelt worked with other referendums, a growth cap, and a new plan for the City of Boulder that used the language of nature conservation to block housing and infrastructure development.

There is no evidence that the Boulder planning board had any intention other than to conserve the city's riverbeds and wetlands. But the greenbelt came into being in a broader context of anxieties about demographics, race, and class and was incorporated into those social relations. In her work on race and landscape in Boulder, the geographer Abby Hickcox argues that while Boulder's greenbelt is often hailed as a first in US urbanism, it belongs to a longer history of using nature conservation in the area to justify expulsion and eviction of first Indigenous and later non-white residents (Hickcox 2007). In addition, Boulder's greenbelt program was a success, garnering the support of the city's property owners because it increased property values while providing alternative explanations for the city's history of settler colonialism, anti-Black racism, and anti-immigration attitudes.

Hickcox reports from her ethnographic work that many people who lived in Boulder in the early 2000s traced the city's lack of racial and class diversity "not to the Klan presence, hostility to working-class populations, or the town's [settler colonial] founding, but to the land use and zoning policies implemented since the 1960s" (2007, 252). As Hickcox goes on: "This version of Boulder's class history reinforces the idea that the virtuous goals of natural landscape preservation inadvertently caused the city to be dominated by wealthy whites rather than vice-versa" (2007, 252). In other words, to borrow from Katherine McKittrick's work on the erasure of non-white histories and spaces in Canada, once brought into the world, Boulder's conservation policies did more than protect green space. They had the effect of naturalizing the long history of erasure of non-white people and places from Boulder by reframing displacement as the unintentional outcome of otherwise benevolent city management policies rather than as processes of dispossession (see McKittrick 2006, 96).

It was also during the late 1960s that Worthington participated in a project to convert crisis-ridden public space into thriving real estate. At the time, Boulder's downtown, like so many other US downtowns, was experiencing capital flight. Worthington—whose office was on Boulder's then decaying main street—and the design team he worked with offered a simple solution to this problem: attract people and their money back to Boulder's central business district by building a mall in the center of the city.

In the Pearl Street Mall, Worthington helped design a place-based leisure experience for tourists that mimicked a bustling public street. It was outdoors. There were sidewalk cafés and bars. There were licensed street performers. There was plenty of sidewalk furniture to sit and people watch. But Pearl Street was no longer technically a public street. It was transferred to a private real estate corporation, which then could determine who belonged on the street (business owners and tourists) and who did not (skateboarders and panhandlers).

There is no evidence that Worthington himself had any intention for the greenbelt and Pearl Street to also be sites of exclusion. But the success of Boulder's greenbelt and the Pearl Street Mall did set the theme for the rest of

Worthington's career. In the late twentieth century, demand was rising throughout the world for urban planners and architects who knew how to transform the ruins of modernist dreamworlds—postindustrial business districts, ports, and crumbling downtowns—into privately owned visitor-class experiences that were also "green" (see Greenberg 2008 on New York City; Eisinger 2000 on Detroit; Rutheiser 1996 on Atlanta). Worthington's next project, the Denver Tech Center, became one of the first private tech city hubs in the United States; again, it was based on developing private urban real estate and using green space to attract private investment. By the turn of the century, Worthington had joined a global circuit of architects and designers who worked internationally. He was contracted to convert crumbling industrial ports in a series of Japanese secondary cities into ecologically themed waterfront promenades. He received contracts to do the same in China, South Africa, and then (after Kigali) Kenya's yet-to-be-realized Konza Technopolis.

Worthington's career is not exceptional, nor is it based on any sovereign "model" of urban design that belongs to any particular geography. His career coincides with a general shift in city management that emerged everywhere throughout the world in the mid and late twentieth century. Modernist styles of building cities, using public resources to provide schools, housing, and social safety nets, gave way to what David Harvey (1989) calls "entrepreneurialism," selling the city's public resources piecemeal to private developers through public-private partnerships while redesigning the city for a visitor class. But the rise of entrepreneurial cities did more than transform urban space and privatize urban management. It created a generation of urban managers, architects, and city branders throughout the world whose job it is to convert the aftermath of crises—fiscal crashes, ecological disasters, and political catastrophes—into investment opportunities.

In its 1960s and 1970s iterations, entrepreneurialism was presented as an austere response to the fiscal crises brought on by heavy public spending and the failure of modernist planning (see Greenberg 2008). In the United States, the "urban crisis," a phrase coined in Eisenhower's 1968 speech "The Crisis of the Cities," was publicly narrated as a housing, service, and infrastructure crisis—discursive work that depoliticized changes that were actually the outcome of anti-Black policing and racism as public policy (see Taylor 2019 and Sugrue 1996 for excellent histories of the urban crisis in the United States).

The entrepreneurial city was about addressing the problems of capital flight by bringing visitors and their money back to the urban core. Public-private partnerships and tax incentives were used to convert public resources and spaces into private real estate projects while city managers redirected finances for everything from public schools and safety to transportation infrastructure toward the production and maintenance of enclaved, theme-based visitor experience.

The shift is self-sustaining. Public-private partnerships and tax incentives work to keep surplus capital from being distributed to social safety nets, public services, and infrastructure maintenance in neighborhoods outside of rebranded enclaves (see N. Smith [1996] 2005). These conditions do not always manage risk and often make the city more risk prone, setting the stage for the next predictable hazard—a market crash, hurricane, major political catastrophe, whatever—to become the next full-blown crisis.

Kevin Gotham and Miriam Greenberg (2014) call this "crisis-driven urbanization." Using New York City and New Orleans as comparative examples, they show how entrepreneurial urban managers cultivated the conditions that would make ecological, economic, and infrastructural breakdowns into full-blown catastrophes that would then require more entrepreneurial reforms. Entrepreneurial reformers in the 1970s created the conditions for events in the twenty-first century—9/11, Hurricane Katrina, and the 2008 financial crisis—to morph from hazards that would have been manageable, had social safety nets and investments in infrastructure maintenance and repair been in place, into catastrophic failures. The next cohort of urban managers in New York and New Orleans converted those entrepreneurially induced crises into opportunities to further raid public resources, setting the stage for another predictable hazard to become the next catastrophe and so on.

Some have called these late twentieth-century rebuilding efforts "disaster capitalism," a concept that draws attention to the ethics of Chicago School economists like Milton Friedman, who famously celebrated Hurricane Katrina as an opportunity to finish the job that neoliberal reforms in New Orleans had begun (Klein 2007, 5). It should be noted, however, that OZ's work in Kigali began before Hurricane Katrina devastated New Orleans. In her 2024 work on Kigali, Shakirah Hudani, influenced by Klein, catalogues some of the green growth institutions that have set up shop in Kigali since the 2008 fiscal crisis as "crisis capital." However, in both Klein's conception of disaster capitalism and Hudani's derivative crisis capital, there is a division of labor between catastrophe and capitalists in which the former makes the disaster and the latter comes in after the fact to take advantage of it. But when the agents of sustainable urbanism—entrepreneurial city building, incentivized urbanism, city branding—are placed in their correct chronological order, a different picture comes into view: one in which crises are intentionally cultivated and marketed as raw material to build new cities, rather than preexisting conditions.

Kigali, of course, is not a postindustrial city. But when the government of Rwanda hired OZ Architecture, it was doing more than bringing foreign consultants to Kigali to draw lines on a map. They were aligning the Rwandan capital to a specific genealogy of planning and financing that works by converting crises—of infrastructure, ecology, and economy—into unique investment opportunities. It is a process that requires a set of institutions and policies that are, like CorpGov in Huchu's "The Sale," designed to facilitate the needs of corporations and visitors while disavowing what is already there.

Kigali was also different from 1960s Boulder and post-Katrina New Orleans in another key regard: the crises that structured the mental image of Kigali in the global imagination had already been resolved when the master planning process started. By late 2007, when the City of Kigali, OZ, and several US journalists began promoting Kigali as a postcrisis tabula rasa waiting to be reinvented, Kigali no longer needed to be rebuilt or reinvented. Rwandans had already rebuilt their own city.

ASSEMBLING THE PLAN

They descended on us! The Caterpillars [bulldozers] started over at Sopetrad [a petrol station] and pushed through [the neighborhood] destroying our houses . . .
—RECORDED INTERVIEW, MARCH 6, 2015

The goal of this plan is not to control the growth of Kigali in an authoritarian way but to provide the framework and flexibility to instigate a long-term process that is guided by Rwandans as they shape their capital city incrementally, responding to changing needs; and that creates a vibrant, unique and truly Rwandan city.
—OZ ARCHITECTURE 2007, XIV

We were still in our homes when they started destroying the neighborhood . . . [there was] no warning. We had to grab what we could and run, right [sibyo, isn't it]? But I have five children! How could I take anything? I had to find my children, get out and run away. I lost everything.
—RECORDED INTERVIEW, MARCH 22, 2015

This planning process has been highly participatory. . . . Through a series of meetings, presentations and conferences with a variety of government officials, stakeholders and citizens, a plan emerged that responds to their needs, concerns, and visions for the future.
—OZ ARCHITECTURE 2007, XVI

You can see that they do not want lower classes [rubanda rugufi] *to live in the city [laughter].*

They want like what is in the plan [igishushanyo, design], *you know? Like those trendy houses for white people* [amazu agezweho y'abazungu] *[more laughter].*
—RECORDED INTERVIEW, FEBRUARY 13, 2015

It is hoped that the Kigali Conceptual Master Plan will help to create a city that is . . .

A livable community that supports a sense of belonging, cooperation and healing, and optimism.
—OZ ARCHITECTURE 2007, XIII

In 2005, three years before the demolitions began and two years after Kagame met Worthington in Aspen, the government of Rwanda hired OZ Architecture. The firm assembled a team of architects, engineers, and designers to write the Kigali Conceptual Master Plan.

"A lot of us architects," Worthington would say later, "we are the musicians, but nobody ever writes the symphony, nobody ever writes the scores . . . what's the big theme? . . . So most of my passion has really been the bigger scale . . . putting the parts together *and then* get all the architects—tell them it's Stravinsky, or Brahms, or *something* but don't just let it all be like Los Angeles where it all falls out of the sky" (City of Boulder 2013).

The landscape architect and geographer Donna Rubinoff was selected as the team's sustainable development expert. Rubinoff brought with her several years of experience working on development issues in the Global South. "We are really very conscious," Rubinoff told a US journalist in 2008, "that everything we do is about trying to transfer the skills and capacity for building to them [Rwandans]" (Marshall 2008, 26). Rubinoff would go on to work for the City of Kigali for several years. In her written work, Rubinoff expresses a commitment to the United Nations' definition of sustainable urbanism as "growth without compromising the precious resource base" (see Rubinoff 2014).

Members from the Colorado chapter of Engineers Without Borders USA were brought in to design housing typologies (see chapter 4). The organization works at the micro level and brings a "bottom of the pyramid" approach to design and development. Assume, for example, that the city's water service delivery network is incomplete. One approach would be to address unequally distributed services by building a new hydraulic system with public funds. Another approach—Engineers Without Borders' approach—is to design a housing unit that circumvents the network altogether with simple and inexpensive technology—for instance, a rainwater collector that can be retrofitted into the house. This approach requires no public funds, and it converts the problem caused by unequal distribution of services into an opportunity, a product that can be sold to city residents—a win-win for cash-strapped policymakers and city residents (see chapter 4 in this book; see also Roy 2010b; Collier et al. 2017).

The OZ team carried out six months of research in Kigali. It presented its findings to the government of Rwanda in 2006, and the product, the Kigali Conceptual Master Plan, was approved by the City of Kigali in November 2007. The Kigali Conceptual Master Plan comprises 198 pages of illustrations and text—all in English.[2]

The ruling party of Rwanda, represented by the mayor's office, aimed to use the plan to attract foreign direct investment by changing Kigali's image: "Kigali," the plan states as a goal, "will be seen as a safe haven that will attract visitors and investment" (OZ Architecture 2007, xv). The OZ team understood that to achieve this goal, it had to enroll crucial "intermediary actors," or third-party promoters to legitimate the plan's claims. These intermediaries included the United Nations divisions, architectural award commissions, and journalists on the Africa rising beat. Thus, another goal for the plan was to "help achieve the [UN] Millen[n]ium Development Goals and a broad vision of Kigali as the 'hub' of the 'new Africa'" (OZ Architecture 2007, xv). Then, anticipating Gregg's words by thirteen years, the plan promises that Kigali will become "the most livable and sustainable city in Africa" (OZ Architecture 2007, xv).

Because of the different stakes—financing and promotion, wetland conservation, humanitarian design, and control over the city's development—there is no single plan in the Kigali Conceptual Master Plan. In chapter 1, "The Master Plan Process," for example, the OZ team references "community participation" as a key method for accessing information and developing proposals. Ample visual evidence of "the community" participating populates this chapter. There are images of Rwandans pointing to Venn diagrams on poster boards. According to a workflow chart at the beginning of the chapter, the data from these community participation events were captured in chapter 2, "Physical, Economic, and Institutional Contexts." But none of this data carries over to actual policy. Instead, much of the data OZ uses to describe the city's existing conditions comes from secondary sources—preliminary UN-Habitat reports based on data compiled in the early 2000s that conflict with the participatory reports.

The OZ team also suggests sustainable solutions to problems whose existence it does not establish. It proposes a bus rapid-transit program without establishing that anything is wrong with the city's popular transportation networks. The OZ plan proposes carbon-cutting measures, but Kigali does not have a significant carbon footprint that needs to be managed.[3] It proposes recycling programs without acknowledging that nearly everything in Kigali was already recycled through existing networks of locally owned businesses and popular economies.

Still, it would be a mistake to dismiss sustainability in the OZ plan, and every plan since, as only a "green veneer." The city's residents and its ecosystems play a key role in how Kigali's future is envisioned in the first plan. For example, Kigali's ecologies—the city's wetlands, forests, watersheds—are described in detail in the OZ plan and every plan since. The OZ team proposes that the government protect land near the mouth of stormwater catchment areas to conserve the wetlands for what they call environmental treatment zones, which are envisioned as a low-cost solution to inadequate infrastructure. In a paragraph that is repeated four times in the master plan, the OZ team writes: "Areas delineated as wetlands or wetland buffers should be protected. Wetlands serve an important ecological, infrastructure, and economic role in Kigali. Ecologically, wetlands reduce erosion and flooding and provide habitat for birds and other plant and animal species. Wetlands are the main component of Environmental Treatment Zones (ETZs) and help to improve water quality and treat wastewater in Kigali" (OZ Architecture 2007, 70).

As "infrastructure," the Environmental Treatment Zones (ETZ) both work to protect the wetlands from human settlement and employ the wetlands as a creative, low-cost solution to an otherwise expensive problem—the city's lack of a central sewage system. And they will do this through what the OZ team calls "multiple, synergistic functions" (OZ Architecture 2007, 56). The synergistic functions are many: In the treatment zone, Rwandan workers will convert waste into useful things, preventing storm and wastewater from polluting the wetlands doing the work that would otherwise be done by a central sewage system. The hypothetical workers will feed effluent into biogas converters, generating sustainable cooking fuel that can then be used or sold, converting waste into value. Others working in the ETZ will sort waste into recyclables. In this way, the ETZ will simultaneously achieve "storm water treatment and flood control; wetland management; wastewater treatment" and become a site of productive activities such as "composting; sorting [and sale] of recyclables; biogas generation and processing; briquetting; staging area for organic waste collection" (OZ Architecture 2007, 56). The people who work in the ETZ will themselves be converted into value both through labor—doing the work of waste management while saving on expensive infrastructures that would normally do that work—and as a new market: consumers for more sustainable commodities like biogas converters.

What is not said in these passages on the ETZ is that many of the "synergistic functions" presented as potential business opportunities were already being performed long before the OZ team arrived (see chapters 2 and 4). What makes the ETZ "sustainable," then, is not the low-carbon activities themselves, but the conversion of activities that have always gone on in the wetlands, such as brick making with local wetland clay, waste collection, and recycling, into growth-based industries that can be captured by investors. Members of the OZ team also played a role in promoting the city of Kigali as an investment opportunity. As the OZ team states in the plan, the City of Kigali would need to provide incentives to global corporations—such as tax-free economic trade zones and discounted land (see OZ Architecture 2007, 105–8). But nearly every city in the region already offered those perks. To make Kigali truly competitive on the world market for foreign investment, they needed to identify Kigali's irreplicable qualities. Those qualities also follow utopian plot points: the end, after the end, and the new beginning.

MARKETING THE POSTCRISIS

In addition to reframing Kigali and its ecologies as opportunities to develop synergy between the environment and economic growth, the OZ team worked to promote Kigali in a way that would be compelling and exciting to an audience of foreign investors. In late 2007, that audience was familiar with a particular narrative about Rwanda—genocide. The OZ team made that legacy part of Kigali's marketing narrative, painting it as a city rising from the ashes after the end.

As a US journalist who interviewed Worthington about the plan during this time writes, "Fourteen years after Rwanda endured one of the most savage genocides in human history—an unfathomable 1 million Tutsis [sic] and moderate Hutus [sic] slaughtered, largely with machetes, in 100 days—the country appears to be outliving the gruesome legacy portrayed in the 2004 film 'Hotel Rwanda'" (Marshall 2008, 24). This prelude to reporting on the plan is not mere description. The journalist begins with the end—"one of the most savage genocides in human history"—and then makes this image familiar by referencing a well-known Hollywood account of the event. Once established, the journalist introduces readers to a new beginning that is being built by a cast of investors from Colorado. "The Rwandan government," the journalist reports, "has since hired Denver/Boulder-based Oz [sic] Architecture to craft Kigali's master plan for the coming half century. Third-generation Denver native Gaylord Layton, founder of Formation Technologies, just opened a luxury tourist lodge . . . Rob Fogler, a Denver attorney, has launched Thousand Hills Venture Fund, expressly for those who want to invest in Rwandan enterprises" (Marshall 2008, 25). Worthington joins the journalist as another liaison between the audience (his society in the United States) and the government of Rwanda. "They are trying to make Rwanda the most sustainable, high-tech, wired country in Africa, a little bit like what Singapore became . . . They [the Rwandan government] are trying to reinvent the entire country" (Marshall 2008, 26).

Worthington repeats this theme in another interview with a different US journalist that same year. "It's more than master planning a capital city," he says, "it's literally writing a script for an entire country" (*5280 Magazine* 2008). In most media releases and promotional reporting, it is Worthington, other foreign consultants, and journalists who do the work of promoting the OZ plan, not Rwandan officials. As the above passages suggest, they are addressing audiences outside Rwanda who—thanks in large part to media images and general narratives about Africa—might be skeptical of the capacity of Rwandans to rebuild their own city. These depictions are not merely media stereotypes about Rwanda and Africa. As members of the target audiences' society who visit Kigali and then report back, the OZ team and the US journalists they enroll play a crucial role in generating utopian energy for investing in Kigali's future. They are the narrators who serve as liaisons "between the reader and the utopian community" (Bedore 2017, 7–8). In the above accounts, the image of Kigali as still in crisis is confirmed rather than challenged, keeping it in place with no reference to how Kigali residents had already rebuilt and repaired their own city.

With the postcrisis tabula rasa status of Kigali established, the OZ team needed to account for the existing conditions of the built environment—the city that Kigali residents had already built based on the historically systematized techniques described in chapter 2. They did this by describing a city that, in addition to being postcrisis, is also *in* crisis.

AFTER THE END

In the OZ plan, there is much tension between the existing conditions of the city, the quality of its organization, and an image of crises that justifies sustainable interventions. For example, chapter 2 of the Kigali Conceptual Master Plan, "Existing Conditions," offers two types of urban neighborhoods in Kigali—"informal settlements" and "sprawl": "While densely populated informal settlements are remarkably sensitive to the topography and provide a good mixed-use framework, they have inadequate infrastructure. In the second type of urban neighborhood, returnees and internal migrants have randomly developed their homes with a 'suburban' quality that tends toward sprawl" (OZ Architecture 2007, 22). This passage recognizes that the city's majority "informal settlements" may need infrastructural upgrades but are also "remarkably sensitive" to the city's ecologies and topography, with a "good mixed-use framework." This observation that majority neighborhoods were structurally sound and already organized in a good framework matches other, independent architectural studies in Kigali (e.g., K. Smith et al. 2018, 48).

On the following page of the OZ plan, however, this nuanced approach to existing housing typologies disappears, and statistics rather than qualitative research are presented as evidence of an urban crisis in the making. "According to available figures," the OZ team writes, citing a 2003 UN-Habitat report, "83% of the urban population of Kigali is located within informal settlements. This number represents approximately 62% of the land area" (OZ Architecture 2007, 23).

The above statistics about informality—unlike the direct observations of neighborhoods that are remarkably sensitive to the city's ecologies—confirms a UN-Habitat (2003, 26) report, *Slums of the World*, that estimates that 82.9 percent of Kigali's population lives in "slums." The following passage proposes organizing the city into housing clusters inspired by Kigali's current built form, derivatives of the city's popular built environments (see OZ Architecture 2007, 32). This tone toward Kigali's "informal settlements" as an inspirational model for the rest of the city shifts later on. If left unchecked, the OZ plan states, "informal" neighborhoods "will continue to tax the existing urban and community infrastructure and availability of developable land within the city boundaries." As the plan states, "The implication [of the UN-Habitat report] is that a large proportion of the city's population is living in highly dense, substandard conditions, with poor infrastructure services. The population increases and continued urbanization trends, if not properly managed [*sic*] will result in increasingly failing environmental and urban systems. The Kigali Conceptual Master Plan is part of the effort to reduce or eliminate this trend" (OZ Architecture 2007, 23–24).

Unlike the passage that understands the city's informal settlements as "remarkably sensitive" to the city's topography and an inspiration for urban clusters, this passage argues that "informal settlements" and the population that constructs them are a threat to the city's ecologies. The city's "increasingly failing environmental urban systems" are, in this view, caused by Kigali residents who apparently lack the skills to build durable structures and are a threat to the environment by producing "highly dense substandard conditions."

Foreign consultants like those on the OZ team are often dismissed as naive because they are assumed to have shallow cultural understanding of the specific environments they are operating in. But much of the team who wrote the first OZ plan had ample experience working in the region and were trained to identify local solutions to development problems. They had a keen interest in the ecological significance of local building technologies and worked to promote these as viable low-cost alternatives to more expensive solutions.

As I show in chapter 4, part of the inconsistent approach to the city's "informal settlements" as both a potential solution and a serious problem in the OZ plan and every plan since was not due to a lack of interest in local building typologies. The planners, engineers, and architects who worked on housing typologies had trouble reconciling the diverse range of housing types within the categories of "informality" and "crises" that they were given to work with as marketing tools.

Their own conviction that the city's "informal settlements" were in fact constructed with durable, locally sourced, low-carbon materials that should be leveraged in organizing the city came into conflict with a potential source of outside revenue: the need to confirm what housing reports claimed: that Kigali, as a post-crisis city, had a housing deficit, and this deficit could be made into a market for newly built affordable homes.

In the above passage, UN-Habitat reports of a shelter crisis in Kigali—that, unlike the OZ team's observations, are based only on secondary statistics—are offered as evidence of a potential market for new construction. "In this way," the OZ team writes, "housing [demand] can promote economic development and in turn generate new housing in the private markets" (2007, 50). In order to make the new markets, city managers would later need to match the built environment to housing market reports and statistics through demolition (see chapters 4 and 5).

THE NEW BEGINNING

Half of the OZ plan proposes conserving the wetlands while converting the existing city into new markets for affordable housing, green commodities, and the expert labor to build and organize environmental treatment zones. The other half is about the plan's big theme: a new city center next to the proposed site of an international airport in Bugesera and thirty kilometers outside the existing urban core.

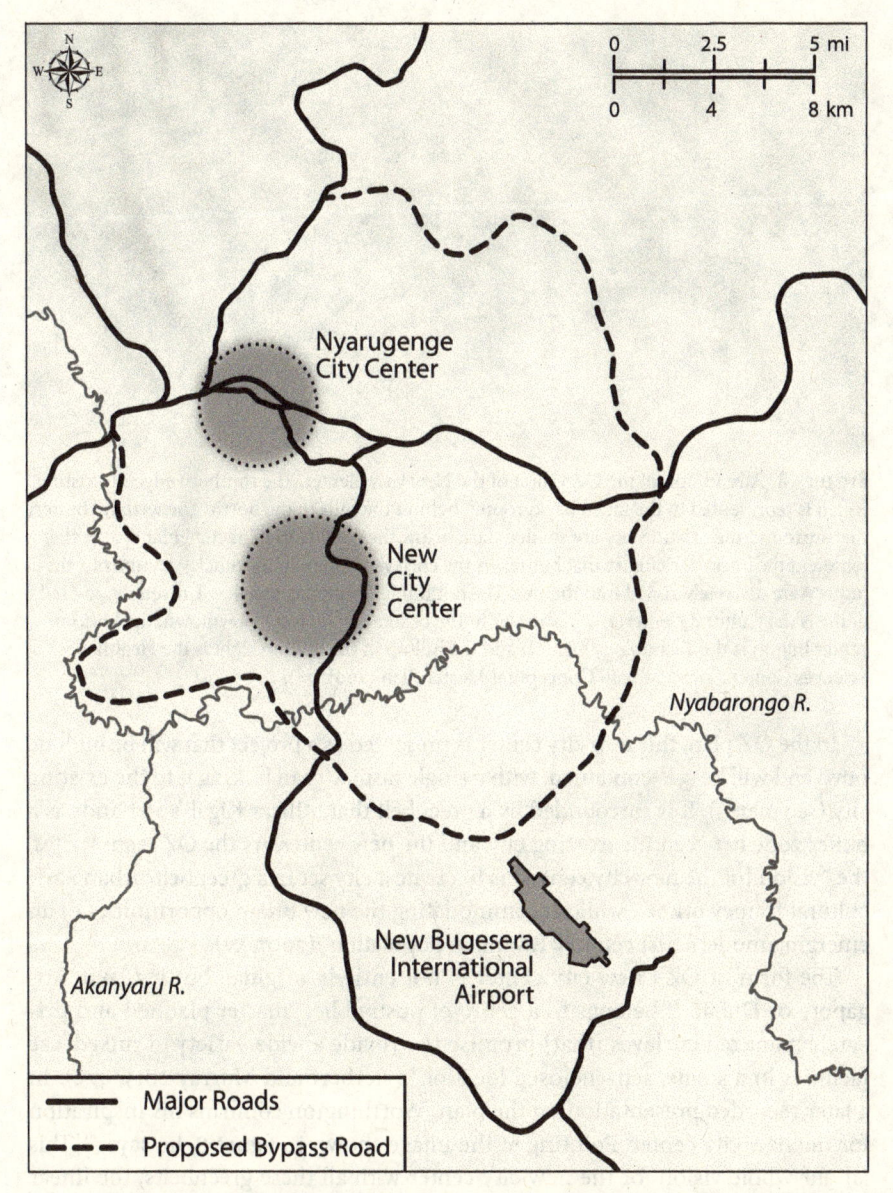

MAP 3. A map representing the OZ team's vision for the location of the New City Center in relation to other locations in the city. Note the location of the existing city of Kigali north of the New City Center, with one proposed road offering a direct link between the two. The wetlands serve as a buffer between both. Source: Kigali Conceptual Master Plan (2007, 75).

FIGURE 8. A rendition of the OZ vision of the New City Center. The southern edge of existing Kigali is represented in the faded background, behind the hills to the north. The wetland buffers that surround the existing city are shaded dark below the patchworked rectangular spaces that represent the linear agriculture that buttresses the enclaved center. In the official documents, the hilltops are discretely zoned into themes. The bottom left hilltop is the New University; the red dome is the Cultural Center; the high-rises in the center are the New Downtown; the bottom center hilltop is the Technology Center; and the hilltop in the bottom right is the Health Sciences Center. Source: Kigali Conceptual Master Plan (2007, 94).

In the OZ plan, this new city center is presented as a project that will be built de novo and will be self-contained, with a single tarmac road linking it to the existing city (see map 3). It is surrounded by a greenbelt that utilizes Kigali's wetlands as a buffer zone between the existing city and the new center. As the OZ team writes, the "vision [of the new city center] is to create a city set in a greenbelt/urban agricultural framework . . . while accommodating the new urban opportunities of an emerging modern 21st century Rwandan population" (2007, xv).

The form of OZ's new city center is not entirely original, but it is not Singapore or Dubai. It belongs to a genre of postpublic, "master planned and privately managed enclaves [that] promise to provide a wide variety of mixed-use facilities in a single, self-enclosed location" (Herbert and Murray 2015, 472). In a later recorded presentation on the plan, Worthington confirms his inspiration for the new city center. Pointing at the image shown in figure 8, he says, "[This is] our whole vision for the new city center with all these greenbelts, the linear agriculture, just like we started here in Boulder—So, Boulder really has kind of been a metaphor [for Kigali]" (City of Boulder 2013). The plan doubles down on that metaphor in a series of photographs of other spaces intended to represent Kigali's future while inviting its target audiences of investors to imagine themselves in that future.

Apparently, the new city center will be inhabited by white subjects in green spaces. One of these images of the new city center frames a streetscape of people sitting at tables, leisurely ordering food, hanging out on sidewalk furniture (see OZ Architecture 2007, 91, for these images); in another, they are riding bicycles through an underpass (see OZ Architecture 2007, 91)—a theme that carries over into later plans. While the OZ plan does not cite the origin of these images, they are from Boulder: the Pearl Street Mall, the civic center, the University of Colorado's underpass.

The idea is that building a new city outside of the existing core will shift Kigali's center of gravity south, away from the current center, Nyarugenge, and toward the international airport at Bugesera. In this vision, visiting tourists, conference-goers, researchers, and entrepreneurs will step off their international flights at Bugesera airport and travel to their hotel, research facility, or business without having to interact with the existing "informal" city. Separated from the rest of Kigali, the new city center is given a sense of place through discretely themed nodes: education, downtown shopping, health and research facilities.

In his analysis of similar depictions of green futures in nearby Kampala, Uganda, Jacob Doherty argues that these images "should be interpreted not merely as a telling lapse in architectural rendering" (2022, 149). Rather, these visions of the future are "consistent with the racialized and racializing norms, expectations, and orientations that structure city making projects" (2022, 149). Drawing on a rich scholarship in postcolonial studies and racial capitalism, Doherty argues that these images disclose the—albeit reconfigured and less overt—long history of colonial and imperial infrastructural residue, even as they are paid for and supported by African governments (see also Appel 2019; Pierre 2012).

In a different study of contemporary large-scale infrastructural projects in Nairobi, Wangui Kimari and Henrik Ernstson show that even when framed in postracial language and including south-south cooperation, contemporary development projects reproduce a subtext of imperial racialization, upholding "long established material and vernacular tropes that frame Africa simultaneously as deficit and the next frontier" (Kimari and Ernstson 2020, 827).

Of course, Kigali, like Kampala and Nairobi, has its own history of attempts to racially segregate the city (see chapter 2). But the new city center, when considered alongside the plans for the existing city, visualizes more than the residue of past projects or efforts to erase Black subjects from a "world class" future. It also demonstrates a vision of how the ideal city might be spatially organized along racialized and classed hierarchies. In the OZ plan, Kigali's majority African residents are not erased. They are very much included and are envisioned as consumers of affordable housing typologies and labor in the environmental treatment zones. This spatial hierarchy between the new city center and the existing city carries over to an investment hierarchy in which the new city center for a visitor class will receive the major infrastructural projects—international airport, health and higher education facilities, leisure green space—while the existing city will be for cheap ecological infrastructures and experiments with privatized affordable housing. In addition, the image of the new city center as outside the existing city, like the discourse of the postconflict tabula rasa being reinvented, offers viewers a vision of urban renewal-as-redemption without any reference to the violence of destruction and displacement that normally underwrites these projects. As Huchu shows in "The Sale," the costs of selling the city are not only hidden by planning images. They are outsourced to humans (and nonhumans) who live in the city as the fictions are made real.

A CITY RULED BY FICTIONS

They did not need to kill the goat.

In 2015 I was with Ms. Muhizi and Ms. Ingabire, two former residents of Kiyovu cy'abakene in a thriving popular neighborhood built by people who had been displaced by the urban core in the first round of master plan demolitions. We were sitting on sofas in Ms. Muhizi's comfortable living room. Our conversation centered around a day in 2008 when the mayor's office had sent police and bulldozers to confront a group of forty-eight families in Kiyovu cy'abakene who were holding out for better compensation for their property. There was a lot of laughter as the women recalled how they and their neighbors reacted. Taking over the conversation, Ms. Ingabire remembered a goat.

In the early morning of July 20, 2008, Ms. Ingabire stood outside her house as she did each day, enjoying the early hours of dawn. As she looked out, she noticed bulldozers and a crowd of young men moving up the road that separated her neighborhood, "poor" Kiyovu (cy'abakene), from its upscale neighbors in "rich" Kiyovu (cy'abakire). The contingent of workers and bulldozers turned off the tarmac road and entered the neighborhood.

That was when Ms. Ingabire noticed the goat. Someone had tethered it to a pole a few meters from her porch. She looked at the goat. The goat looked at her. She looked at the bulldozers and the crowd of day laborers as they advanced closer. Then, she looked back at the goat.

Clouds of dust and debris rose behind the goat.

She never saw the goat again.[4]

That day, the demolition crew took out three generations of homes, businesses, dance halls, bars, a large market, and the common grazing area for goats. Decades of sweat equity, community building, and rebuilding were erased in a day.

Ms. Ingabire's recall sped up. "People started screaming, they were yelling at each other to get out of their houses, you know [*urimokumva*]?" She went on: "They yelled, 'The war is back [*intambara yagurutse*]!' . . . . People beating on pots, or whatever they had to wake each other up. God! That day . . . wow [*mbega we*]!

... It was traumatizing [*baduhahamuye*, they traumatized us]" (recorded interview, March 6, 2015).

According to Ms. Ingabire, some residents tried to stop their homes from being demolished. They threw rocks at bulldozers. They screamed at the police and security guards. Those who protested found themselves sitting on the side of the road under police guard.

"You know about ninety-four?" Ms. Kayagire, a former neighbor of Ms. Ingabire, asked me in a separate interview. "You saw that somewhere [*wayirebye*]?" I mumbled an affirmative. "Okay," she said (using the French, *bon*): "So [*nonese*], if you go through that . . . and then you wake up one day and there are police everywhere; . . . you understand [*wumvise*]? Some went a little crazy [*basaze*, a pejorative for "insane"], [it was like] the war had returned [*intambara yagarutse*]" (recorded interview, March 22, 2015).

Ms. Bizimana, another former resident from Kiyovu cy'abakene, put it like this: "Everything, I mean *everything* was destroyed. [My neighbors] were outside in the streets. . . . They [the police] warned me not to cause trouble . . . There were police everywhere . . . ." (recorded interview, November 23, 2014).

This reenactment of the traumatic past was not only the unfortunate epiphenomenon of past trauma activated by demolition. It was, rather, the first attempt to match the city to its brand image of a postcrisis city waiting to be rebuilt from the embers and to use demolition to resolve the contradictions of finance-driven sustainable urbanism.

Scholars of the Rwandan state have consistently understood the demolitions that followed the OZ plan as a top-down authoritarian project to clear away "the old disorder" to build the new city (see Hudani 2024). But rather than the consistency of top-down and well-planned urban takeovers, the first round of demolitions revealed how much trouble city managers had in matching the existing city with its brand image while still trying to govern it. In attempting to create the tabula rasa, the demolitions were making urban life more difficult to manage as city authorities became ruled by the fictions of their own creation.

"The idea is to restrict movement . . . to restrict development to low-lying areas. I don't want people moving up Mount Kigali or Mount Jali where there is no infrastructure . . . ." On February 21, 2014, I sat across from an architect and urban designer who worked for the Singaporean firm Surbana-Jurong at Kigali's One Stop Center. In 2009 the City of Kigali hired Surbana-Jurong to convert OZ's conceptual plan into a detailed, comprehensive master plan. Surbana-Jurong embarked on a process of mapping the city and writing laws, building institutions in an effort to create the foundation for implementing OZ's plan.

The consultant that day had invited me to attend a training session on the zoning database for engineers who work for the Ministry of Local Government in sector offices. The objective of that day's training was to demonstrate how the zoning plan worked with the new online permit and building registration system. To do the work of planning and institution building, Surbana leveraged this expanding state bureaucracy. For example, thanks to conflicting colonial and postcolonial land tenure regimes that made it impossible for most Kigali residents to formally own land in the city, knowledge about who owned what land was restricted to neighborhood-level leaders. The government of Rwanda addressed this problem by bringing in a UK consultancy to build a new land tenure system without recognizing the existence of previous land regimes (discussed in chapter 2).

From 2007 to 2009, the government of Rwanda hired a UK-based private development firm to run a pilot land registration program in Kigali, funded mostly by the UK Department of Foreign International Development, that would provide the first comprehensive map of every plot in the city. To do that work, the consultancy built more bureaucracy, setting up land registration centers in all thirty-five sectors of Kigali while operating as if there were no other land regimes working in the city.

Thanks to the land registration program, every recorded parcel in Kigali was numbered and compiled into a single database that was then converted into a digital map. Surbana then took over that bureaucratic tool, embedding itself in the power structure of the city of Kigali while transforming the Rwandan state. The land tenure system, built by a UK firm and financed by the governments of the United Kingdom and the Netherlands, led to a GIS database built by Surbana. The designer I met with was from Europe, and he worked for a Singaporean firm. He worked at the city's One Stop Center, an office run by the city. Sector-level engineers who worked for the Ministry of Local Government reported to him. And he reported up to a US planner who was originally on the OZ team but then transitioned to senior advisor to the City of Kigali. In other words, OZ and Surbana, along with the UK consultancy who wrote the new land regime, were very much part of the city's government.

Over the next four years, Surbana built out OZ's five-chapter conceptual plan into a multivolume Kigali City Master Plan and a Land Use Master Plan. From there, Surbana used the digital mapping program ArcGIS to generate an online, interactive map, using single-use coding to show what *should* be built over the existing city. All the master plan information was then made available to the world, published on an open-access website.

The document trail for Surbana's 2013 master plan is impressive: three detailed subarea master plans totaling more than seven hundred pages, a land use master plan, a transportation master plan, two years of PowerPoint presentations from multiple training sessions with Kigali authorities, and video clips posted to social media sites to grab the attention of CNN and NBC (see also Shearer 2015). Surbana mapped and coded each individual parcel and the location of every neighborhood that was in line with the plan. By February 2014, when I showed up at the One Stop Center, it seemed all the mechanisms were in place to keep people from "moving around."

At the One Stop Center that day, the Surbana consultant walked me through how to read the plan from the perspective of an engineer deciding whether to approve a parcel division or building permit, or to demolish a building. "The technician logs in," he explained. As he did so, a satellite base map of Kigali's built environment appeared on the conference room screen. On a menu to the side of the map was a list of layers a user could import and lay over the map. The consultant clicked on the "parcels" layer, and every parcel that had been registered in Kigali, complete with its unique number, was marked on the map. He then added the proposed zoning plan and transportation plan over these layers. Technicians reviewing a construction permit would compare the permit requested with these individual components of the master plan to decide whether to approve it.

In his demonstration, the consultant zoomed in on a plot in the Muhima neighborhood, next to where Kiyovu cy'abakene had once stood. With a click, the land use zoning map appeared. He then brought in another layer, the green wetlands buffer. We were looking at an area that had been rezoned in red as C4, a city commercial district for "large-scale commercial development,"[5] meaning, he explained, that the only future construction allowed in that area would be high-rise shopping centers. What was already there—clusters of low-rise houses and shops on small plots—would be (in theory) demolished if a permit to build a shopping mall in the area was requested. As long as the plan remained isolated from any of the contaminating elements that could disrupt its base map—illicit building, infrastructural tinkering, unpermitted repairs, "people moving around"—it was compelling.

"Does it work?" I asked.

"No." He laughed. "It doesn't work; it is not going to work" (unrecorded interview, reconstructed from field notes, February 21, 2014).

According to the consultant, when the government of Rwanda hired Surbana, OZ's plan to build a new city center outside of Kigali needed to be revised. "It was a problem," he explained, "of financing and infrastructure." No budget existed to install a new infrastructural base outside of the existing core—at least not for the people who were intended to use the center, whom the consultant called "an international class." In its plan, Surbana-Jurong moved the new city center back to the original city and designed it over the "informal" neighborhoods that surrounded it. The issue with the demolitions, the consultant explained, was how they were also putting a strain on city services elsewhere. The "informal" neighborhoods, despite being labeled as informal, were fully serviced with water and electricity, making them cheaper to redevelop but creating new issues from displacement.

"People," the Surbana consultant observed, "are now being forced out and moving out to where they cannot be serviced [by urban infrastructure]" (unrecorded interview, February 27, 2014).

Since 2012, in theory (and by law), anytime a structure in Kigali is built or repaired, and anytime land is sold, the developer or seller must first request a permit at a district-level One Stop Center. The permit should be denied or approved depending on what the reviewing technician finds in Surbana's zoning database and land use codes. But, as the consultant pointed out, contradicting an extensive scholarship on state control in Rwanda, "There are plans, we have lots of plans, but there's no political will to implement them" (unrecorded interview, February 27, 2014).

Halfway up Mount Kigali, where the consultant did not want anyone to go, several families explained why they had ignored the regulations. For example, Ms. Mukamana bought a parcel from a landowner in an area initially zoned for passive recreation. "When we first came to the city," she said, "we were renting near the road." Originally, she and her husband were renting a house downhill in a neighborhood that was serviced with water and electricity. "We decided to build a house, you know? To live well [*kubaho neza*]." Ms. Mukamana was unemployed. Her husband was an essential but intermittent worker—a porter in Nyabugogo transportation hub. They had a child, and making rent was becoming impossible. Buying a small parcel to build their own house was a solution to unpredictable wages. They built a one-bedroom house with a living room, an outside kitchen, and a bathroom in a courtyard.

"Do you have your title [*ibyagombwa by'ubutaka*, official documents]?" I asked her.

"No, we haven't seen it," she said.

I asked, "How do you know it is your land?"

Because, she said, "We bought it from the family who owns the title."

Ms. Mukamana's home was on one of eighteen smaller parcels that belonged to a larger plot, owned by one family who had registered their land during the registration drive. As far as the official records are concerned, these titles belong to the same plot. Unable to get official approval for their land purchase—because it was in an area that was coded for ecotourism, not building homes—the Mukamana family purchased their land using previous methods described in chapter 2. They gathered three families and a community leader and drew up their own title, paying everyone inzoga z'abagabo.

No land title, Ms. Mukamana explained, meant no building permit. Worse, applying for a permit would invite a sector-level engineer to look at their property, and the engineer would be able to see what the Surbana consultant showed me: the plot was in an area that was zoned for recreational green space, not for building. In addition, Ms. Mukamana did not have the resources to meet the stringent new building codes. Like most of the city, she had to work with earth and clay, which were, at that time, out of code.

"We had to build in two days!" Ms. Mukamana exclaimed, laughing, "because that is all we could pay for!" (recorded interview, February 26, 2015).

To build outside the reach of the plan, Ms. Mukamana worked with some government authorities, just not the offices that the consultant needed her to. In addition to inzoga z'abagabo, Ms. Mukamana now had to pay what she jokingly called *inzoga z' abayobozi* ("beer for the leaders"). She paid community-policing units officially tasked with keeping the neighborhood under surveillance to *not* do their job and overlook her construction project. She paid ruling party leaders at the *umudugudu* (community) and *agatare* (cell) levels to not see that her house was going up and to remember that it had always been there after it was finished. Two days of this "beer" added up to RWF 200,000 (about $275 in 2014)—more than the actual labor costs for two days of building, which came out to RWF 120,000 for three abafundi and several day laborers (about $171 in 2014).

Ms. Mukamana did not see her house as resistance or defiance to any authority in Kigali. She wanted (as did many of her neighbors) to be legible to Surbana's maps and to have official documents. Being seen by the state was better insurance against being evicted without compensation. And the bribes only made building more expensive, which many people in the neighborhood blamed on the new regulations in the master plan. In paying different authorities to build, however, Ms. Mukamana and her neighbors also invested and strengthened the authority of lower-level state officials over the central government at the One Stop Center. Instead of preventing people from building outside of the plan, the City of Kigali's building codes and Surbana's zoning codes incentivized residents to sidestep the rules by engaging lower-level authorities against the plan. This transferred the authority to determine who built what *away* from planners and consultants at the One Stop Center and toward lower-level authorities, generating intractable problems for technocrats who work on planning.

As my February 2014 meeting with the Surbana consultant went on, it seemed clear why upper-level city authorities were losing ownership of the tools they had designed to control the city's growth. Before they could address how residents like Ms. Mukamana acquire land and build houses outside of city plans and codes, they would need to acknowledge that alternative processes existed in the first place. And that was impossible, without contradicting an element of Kigali's brand image: Kigali is a "better" destination for investment because there is zero corruption in the city (see Crisafulli and Redmond 2012; Ruxin 2013, for examples that boost this claim; Goodfellow 2022, 150, for a critical assessment). When I asked the consultant if he thought bribery might be involved in disrupting his work, rather than the lack of political will, he responded lightheartedly. "I wouldn't talk about corruption," he warned, "if I were you."[6]

This case illustrates not only how difficult it is for city managers to implement plans, but also how diffuse and disconnected urban authority is in Kigali because of the master plan. Before the OZ plan was implemented, its big theme—the new city center—was discarded. The plan needed to be revised and rewritten because of the limits of financing and infrastructure. The demolitions of Kiyovu cy'abakene (described above) had nothing to do with making the city governable: Kiyovu was already enrolled in state institutions, its plots mapped and governed by the ruling party of Rwanda (see chapter 2). The demolition was about aligning the city with its fictive tabula rasa brand image and experimenting with new housing (see chapter 4).

The demolition of Kiyovu cy'abakene certainly disrupted the lives of people like Ms. Ingabire. But it also, according to the Surbana consultant, disrupted the work of planning. The demolitions were activating alternative land transfer and building processes as residents like Ms. Mukamana needed to revert to previous techniques of acquiring land and building. These processes were supported by lower-level authorities and community police units who were all carrying out other state functions that managers at the One Stop Center could not acknowledge without betraying another brand fiction: Kigali was a city with zero corruption. In other words, the biggest disrupter of actual planning work was not a lack of "political will." It was the marketing fictions in the Kigali City Master Plan that had come to rule over the city. The planning challenges that were, ironically, being generated by the plan's utopian images did not, however, interrupt the processes of selling the city to outside investors.

THE PROMOTIONAL STATE

Surbana's zoning database was just the first step in attracting foreign investment to green the city of Kigali. Another layer of state institutions would need to be built to facilitate foreign investment. In 2008 Kigali's management team, which included some of the original members of the OZ team, started sending out calls for investment in high-profile, hype-generating projects. That effort was supported by highly symbolic, widely publicized efforts to position the city as the green alternative to other cities in the region. Just after the plan was complete, for instance, the City of Kigali announced that the Rwandan capital would be the first city in Africa to ban plastic bags. The plastic ban cost very little but returned ample media hype, largely in the form of feature pieces in international media (see Kardish 2014).

To create a safe environment for investment, the city needed to build public-private partnerships, a single institution to aggressively manage and control Kigali's image. They needed systems and laws to facilitate the entrée of foreign firms into Kigali and ensure their ability to access land and consumer markets without red tape or political opposition. The most pivotal move in this direction was an internationally coordinated institution-building project that involved Tony Blair, who became an advisor to the government of Rwanda in February 2008.[7] One of the first projects from this partnership was the Rwandan Development Board (RDB), a corporation-government hybrid designed to serve as a one-stop public-private partnership, providing foreign investors with access to Rwandan consumer and housing markets.

The RDB opened in late 2008. The institution devised a fast-tracked online business registration process, making it possible for foreign construction and real estate firms to register a business and begin operating on the ground in Kigali in just two business days.

Thanks to these reforms, the World Bank ranked Rwanda second in Africa in its 2009 "Ease of Doing Business" Report—a jump of seventy-six places from the previous year. As Kigali rose to the top of the World Bank charts, more government institutions and legal codes were established to police the city's brand image.

As the demolitions continued, the master plan won international awards, including the 2008 United Nations Habitat Scroll of Honor Award, the 2009 American Planning Association Award for Best Comprehensive Plan, the 2011 American Planning Award for Best International Project, awards from the American Institute of Architects and American Society of Landscape Architects, and the 2013 Singapore Institute of Planners Award for Best Planning Project (see Rubinoff 2014 for a full list). And while the plan's brand image was disrupting daily management and other state projects, it has nevertheless been effective at generating the utopian energy that Worthington promised Kagame back in 2003: a unique, special city, with its own allure, as described in the opening vignette of this chapter—a new city for the end of the world.

Making Kigali's brand image involved much more than a national state wishing to demolish a city into a high modernist vision. It included global design firms like OZ and Surbana, parastatal organizations like the UK-based development corporation that wrote the land law, UN-Habitat statistics on housing deficits, and globally circulating images of Kigali as a postcrisis city that was also in crisis. It was driven, without a doubt, in part by some city authorities wishing to change the image of Kigali, but also by ideas about how to manage urban environments in ways that generate new markets. The difficulties in enforcing sustainable policy in Kigali, in acknowledging the existing city, in making a city that is not a tabula rasa look like one, and in generating the synergy between nature conservation and market growth reveal not only how far-flung the project of sustainable urbanism is, but also how heterogeneous and conflicted the state that is assumed to govern it is. The processes of destroying the urban core so it can be rebuilt for the end of the world pitted the "extrastatecraft" (Easterling 2014) networks of international expertise and ministry-level authorities against the continuum of local-level authorities and institutions of land transactions that predate the RPF and the Kigali City Master Plan.

. . .

Ten years after work on the OZ plan began and after the demolitions of Kiyovu cy'abakene, Kimicanga, and Gacuriro were complete, after the Rwandan Development Board was up and running, and while much of the popular city stood in ruins, Sehoon Oh, the former mayor of Seoul turned consultant, declared in a press statement from the Rwandan capital that the city had finally been returned to its tabula rasa status: "Kigali," he stated ". . . is like a blank canvas. It has so much potential in the future. It is possible to make whatever you want of it" (Mwai 2015).

All that was left to do was to make the fictions real.

4

Destruction

Making Fictions Real

Madame Presidente of the Parliament of the Republic of Rwanda
Batsinda, 16 February 2010
Memorandum: Reparations[1]

Madame Presidente,

We are writing to inform you of our unjust treatment by the City of Kigali and the Rwanda Social Security Board. These injustices and the suffering they have caused us are listed below:

1. We are residents from ubumwe cell, also known as Kiyovu cy'abakene, who were relocated to a village in Batsinda. In meetings with city authorities, we have explained the difficulties that this displacement has caused, and that we were moved by force. The City of Kigali agreed that they would work on these problems, but nothing has been done. These are our complaints:
    - The City of Kigali originally agreed that we would pay for our homes in Batsinda over the course of 20 years. However, we were reimbursed very little for our homes in Kiyovu and we do not have any other means to earn money and pay for the homes that we were relocated to.
    - When we were relocated, we were told that it was for our benefit and the benefit of the greater people, but our lives have been made worse.

- The city agreed to give us Rwf 500,000 [approximately $550] to help us start our new lives [. . .] but never actually paid us.
- We relocated as the city told us to, and some of us were paid for our property, but some of us have never been reimbursed for our property, homes, or possessions.
- The city of Kigali said they would renegotiate the compensation for some of us, but they have not done this.

2. In addition to the problems listed above, we have incurred further hardships after our relocation:
    - We had a meeting with RSSB [the national pension fund] on 12/2/2010. They told us we needed to pay back the loans for our houses in Batsinda immediately, ignoring our agreement with the city of Kigali.
    - Some of us cannot even pay for our children's school fees.
    - We cannot find work because we have been relocated so far away from any work site . . . and there are no work opportunities in this area.
    - We are slowly becoming poorer and these new homes are in disrepair because we do not have titles to our homes. Without the titles, we do not have the right to make repairs or to take out a loan. [. . .]
    - The Rwf 500,000 that we were promised but never arrived was supposed to be spent on fixing and finishing our homes when we were relocated, to put in electricity and gardens, build enclosures, and other things to make our homes livable.
    - We have had to use any other money we could find to do these things. The city agreed to help us finish our houses but did not do anything, leaving us here at a great loss.
    - Of those who have not been compensated yet, one of us lost all of his possessions . . . He is now dead, and the City of Kigali still has his compensation check.
    - The others are extremely poor; they survive on handouts from their neighbors.

Madame Presidente,
Before we were displaced, we were told by the City of Kigali that we were going to live somewhere else . . . Those of us who were picked [by the city] to go to Batsinda asked the City of Kigali to agree to at least make the area livable. At the time, they had not finished the homes; there was no electricity, no water. However, when the city wanted us to move, these problems, and the problems of compensation, were still unresolved. They bulldozed our homes, forced us into a bus, and made us all move at once. All our possessions were destroyed.

Madame Presidente,
We request that you help us seek reparations for these injustices caused by the City of Kigali. We can come to parliament, or you can visit us here and we can show you the injustices that we have had to live with for a long time.
Please have a good day at work,
[Signed,][2]

The same year the OZ team finished the master plan, the City of Kigali and the Rwandan national pension fund were sitting on more than two hundred empty houses and facing a deficit of more than $1 million. Built on what was then the edge of the city but is now a urban center, the Batsinda Housing Estate was intended to be a solution to the crisis of "informality" and a housing shortage identified by the United Nations and the government of Rwanda and inscribed in the plan. The estate was, in the words of a German planner and architect who worked on the project, a "model of sustainability and affordable housing" (Ilberg 2008, 7). Each house in the estate was designed to get its water from its own rainwater catcher, its electricity from a private photovoltaic system, and its cooking fuel from a biogas converter. Each newly built, single-family, two-bedroom house could be purchased for RWF 3.5 million (about $6,000 at 2007 exchange rates). For those who could not make a one-time payment, the (now dissolved) Rwanda Housing Bank offered twenty-year interest-free mortgages.

There was only one problem with the Batsinda houses: no one wanted one. The houses sat empty until landowners from Kiyovu cy'abakene were forcibly relocated to them. The lack of demand for these models of affordable housing, built with all the best practices and cutting-edge technologies of sustainable urbanism, was surprising given the apparent scarcity of quality housing in Kigali—according to UN-Habitat (2003) and the OZ plan, more than 80 percent of the city lacked adequate housing and infrastructure, and more than 90 percent of the city's structures were out of code.

Batsinda was intended to provide a sustainable solution to what the OZ team of experts identified as Kigali's shelter crisis, and to do it in a way that was green, low cost, and socially conscious while accelerating the growth of the city's affordable housing market. The architect and the engineer who designed the house saw their work as advocating for city residents by making a case to legalize earth as a building material.

Instead, as the Kiyovu landowners' letter to Rwanda's parliament attests, the Batsinda Housing Estate became a catalyst for destruction and displacement. When landowners in Kiyovu cy'abakene refused to embrace the solution to a housing problem they had already solved for themselves, the City of Kigali turned to extraeconomic force and dispossession.

This chapter argues that the destruction of built environments and consequent production of scarcity are mobilized to make real the planning fictions that drive green capitalism in Kigali. It is about how Kiyovu cy'abakene—a real place that was near zero carbon, built with renewable resources, and owned and operated by Kigali landholders—was reimagined as a rhetorical "slum" and then converted into an actual one by force; and it is about how the Batsinda Housing Estate, a sustainable solution to a fictional crisis of inadequate housing in Kigali, became the catalyst to make that crisis real. I show that the destruction of neighborhoods in Kigali is not only the result of local elites wishing to demolish "slums" or "informal settlements" to build "world-class" luxury cities (see Davis 2006; Ghertner 2015; Harms 2016; Hudani 2024). The destruction of neighborhoods in Kigali is also done in the service of making new markets for green commodities while converting already existing, low-impact, low-carbon construction technologies into capital through the alchemy of expertise. To preserve a monopoly over what counts as "sustainable" and to legitimate the plunder of popular technologies by cheapening them, Kigali's international teams of managers must render other nearly identical low-carbon popular technologies of city building *unsustainable*.[3]

As discussed in chapter 3, two very different approaches to sustainable urbanism underwrite the call for investment in Kigali's green future. The first is the image of Kigali as a postcrisis tabula rasa. To market the postcrisis city, the government of Rwanda hires outside firms and develops its own institutions to promote Kigali as a city that, because it is already imagined as a tabula rasa, is an opportunity to build a "better" and "greener" city from scratch. As I show in chapter 3, this approach is rooted in entrepreneurial city-building practices, based on competition with other cities in the region for outside finance, that convert the planetary crises of ecology and economy into new market opportunities for investors. This tabula rasa approach drives the new city center in the OZ plans and the extended central business districts in the Surbana plans—luxe, green enclaves planned for what consultants call an "international class."

Many critics of the Kigali City Master Plan and other plans like it focus only on this tabula rasa approach as an example of "world-class" urbanism driven by elite fantasies that make no provisions for the urban poor (see Watson 2014, 217). But these criticisms ignore how the OZ and Surbana plans proffer an image of Kigali as a city that, in addition to being postcrisis, is also *in* crisis: a city of "slums" and "informal settlements" where the "urban poor" dwell in squalor, and where unregulated urban growth poses a threat to Kigali's wetlands and other ecologies. Unlike the high-tech, visitor-class enclave of the new city center, surrounded by greenbelts and supported by large infrastructural projects like the new international airport, this approach to green capitalism is, in the words of a consultant who worked on the Kigali City Master Plan and the Batsinda Houses, based on developing "low-tech or no-tech," minimal aesthetics, and cheap market-based solutions to problems of housing and infrastructure that leverage "local technologies" (Ilberg 2013, 149).

As Kiyovu landowners expressed in their demand for reparations, they paid twice—once for each utopia. First, through the coerced dispossession of their land, homes, and workplaces (outlined in point no. 1 of the letter) to make real estate available to build the new city center, and second, through the coerced consumption of "affordable" sustainable housing solutions to a housing deficit that did not previously exist (outlined in point no. 2 of the letter).

After situating this chapter in work on the destruction of built environments and the construction and production of scarcity, I examine how sustainable design and statistical projections worked together to engineer a sustainable solution to a housing crisis that did not exist: the Batsinda houses. I then turn to Kiyovu cy'abakene, a neighborhood that came up as an affordable, low-carbon solution to the problems of housing and infrastructure that no one else, until very recently, wanted to solve. Drawing on Cajatan Iheka's ecocritical perspective on what he calls "imperfect" African ecomedia and Tlotlo Tsamaase's short story "Behind Our Irises," I show that while Kiyovu cy'abakene was not perfect, the neighborhood was built and organized through the expert deployment of low-carbon renewable materials. I then turn to the production of scarcity to show that the demolition of Kiyovu cy'abakene was not only the example of state violence that it is often used as (see Manirakiza and Ansoms 2014; Goodfellow 2022, 225; Longman 2017, 180; Sommers 2012, 218). Rather, Kiyovu cy'abakene was destroyed because of an effort to generate effective consumer demand for the Batsinda houses by producing scarcity. Lastly, I follow through to examine the imperfect yet nonetheless effective ways displaced residents from Kiyovu cy'abakene worked with residents evicted from other neighborhoods to repair the popular city from sustainability-induced catastrophe against destruction and displacement.

DESTRUCTION AS THE PRODUCTION OF SCARCITY

The notion that African cities lack adequate, affordable housing and infrastructure has received considerable attention from policy scholars, journalists, the United Nations, and design practitioners. Of the billion "slum dwellers" that UN-Habitat (2023) has identified as living in the world today, 238 million apparently live in African cities. This statistic, coupled with rapid population growth, is often presented as an "unsustainable" disaster that demands immediate attention—a focal point of UN Sustainable Development Goal 11.

Many scholars have deconstructed these stereotypes of African cities as "unmitigated sites of crisis" with nuanced evidence to the contrary (see Iheka 2021; Myers 2011, 2016; Mbembe and Nuttall 2008; Quayson 2014; Robinson 2006; Simone 2004a). In addition, a strain of ethnographies written from African cities demonstrates how neighborhoods that have been coded as "slums" and "informal settlements" are, in fact, sources of environmental activism (Kimari 2022), technological innovation (Guma 2023), and economic ingenuity (Kinyanjui 2014; Mutongi 2017).

Yet, despite so much scholarship that challenges discourses about deficiency in African cities through on-the-ground, empirical evidence to the contrary, "the relentlessly negative and ultimately unbelievable tropes that dominate how much of the world understands cities in Africa" nevertheless persist (Hoffman 2017, 5). This persistence suggests, following Hannah Appel (2019), that the "unbelievable tropes" about African cities in crisis are not just bad ideas to be deconstructed through empirical evidence to the contrary. They are essential to the project of green capitalism itself—specifically, the making of new markets for expert labor and zero-carbon solutions to scarcity of housing and infrastructure. After all, as the Cities Alliance, a partnership between the World Bank and UN-Habitat, points out, if you add up the collective resources of the world's billion slum dwellers, you get a "purchasing power estimated at more than $5 trillion," all of it "represent[ing] an uncaptured market for the formal private sector" (Cities Alliance 2008, 3). That "uncaptured market" is effective demand for sustainable commodities like tiny houses, solar panels, and rainwater catchers—commodities like the Batsinda houses. Issues arise when the projected consumers—the uncaptured market—are not "slum dwellers" waiting for better commodities but already have their own houses.

In the sections that follow, I show that the housing shortage in Kigali is not an issue that consultants find on the ground and then respond to with sustainable solutions. Rather, housing scarcity in Kigali must be "*made* through discursive and material practices" (Anand 2017, 32), or what scholars of water infrastructure aptly call "the production of scarcity" (Barnes 2014). In Kigali, scarcity is produced through technical, economic, and social processes that require the destruction of neighborhoods, the dispossession of knowledge, and the transfer of building technologies from Kigali residents to private firms. Demolition erases what is there but also generates profitable problems to solve. These processes require us to look beyond a vaguely defined "state" as the only source of extraeconomic violence and into the roles that globally circulating narratives about African cities in crisis and humanitarian design solutions to those problems play in setting the stage for demolition and displacement.

This chapter is based on ethnographic research with former Kiyovu cy'abakene residents, including over one hundred recorded interviews with former landowners and tenants that I collected between 2013 and 2015, and my own participation in construction projects. As my research progressed, families from Kiyovu cy'abakene took me on walking tours of their former neighborhood to map key infrastructure, sites of cultural production, and common green space. On my return trips to Kigali, I caught up with the families from Kiyovu cy'abakene whom I know well and documented the changes in Batsinda. What I think I know about the experts who worked on the Batsinda house is limited to what they left behind: the Batsinda houses, a construction manual, the Kigali Conceptual Master Plan, and their own publications. To be clear, my aim is not to condemn environmentally conscious engineers and architects for doing their jobs or to suggest that the consultants who designed the prototype contributed to the City of Kigali's decision to demolish Kiyovu cy'abakene (they didn't) (see Ilberg 2008, 9; Ilberg 2013, 155–56). But if green capitalism is, like capitalism in general, "a project" (Appel 2019), then we can benefit from understanding how green projects are assembled and then brought into the world.

DESIGNING SCARCITY

In 2005 a US engineer and a German architect began designing a prototype affordable two-bedroom house on the outskirts of Kigali. Both consultants worked on the Kigali City Master Plan and for the Rwandan state. The German architect was then an advisor to the City of Kigali and would go on to become a senior advisor to Rwanda's Ministry of Infrastructure. The US engineer worked for Engineers Without Borders, an NGO brought to Kigali by OZ Architecture to work on the Kigali Conceptual Master Plan. The consultants documented how they built their prototype in a 2007 "Low-Cost Construction Manual." They hired Rwandan abafundi (skilled masons who know how to work with local clay), who excavated clay from land near the construction site using hand tools. They then sifted the clay for anomalies, added water, and let it stand for a day. The following day, the abafundi molded the clay into compressed bricks using a hand press. They then laid the bricks out to dry under a plastic tarp. Later, the workers excavated a shallow foundation and made a floor with rock from a nearby quarry. The abafundi then installed a biogas converter just outside the foundation and made a frame with African bamboo (a renewable wood). They installed a rainwater catcher above the biogas converter. They then built four load-bearing walls with the compressed air-dried bricks. The total cost of the house was, according to the manual, "less than 2 million Rwandan Francs" (about $3,000 in 2007 exchange rates) (Ilberg and Rollins 2007, 4).

The house also appeared in the 2007 Kigali Conceptual Master Plan. In the mayor's introduction to the plan, the housing prototype was presented as a solution to insufficient housing supply in the city. "With regard to the housing issue," the mayor writes, "the model house project . . . was used to show that local materials can be used to build decent and safe low-cost housing" (OZ Architecture 2007, ix).

Shortly after the destruction of Kiyovu cy'abakene, one of the consultants who worked on the house published an article to distance their work from the demolitions. The designers' intention, they wrote, was to increase, not decrease, Kigali's housing stock and to legalize earth as a building material. More homes, they reasoned, would lead to more supply; rising supply would eventually equalize with demand, which would cause housing prices and rents to drop in the city, leading "to even more numerous, adequate housing units at lower sales and rent costs" (Ilberg 2008, 8).

The mayor's account about the house in the plan and the consultants' account in their own publications after the demolitions are useful for two reasons. First, they identify a specific problem the house was intended to solve. That problem—a shortage of housing—is, in market terms, the effective demand for more supply. Second, this account provides a clear theory of how the house would solve the problem of scarcity in a sustainable way—more affordable houses built with local materials that have a low impact on the environment will also lower prices, so the poor can afford to rent or buy in the city.

While the consultants who worked on the house present this theory as just how things work, it is a departure from conventional development theory. A conventional approach to poverty and housing scarcity would propose large-scale subsidized housing projects built into the city infrastructure. By contrast, the consultants and the mayor assert that what needs to be corrected is not the housing supply but the market, in which too many people chase too few houses, driving prices beyond the ability of the poor to participate in the market. More houses, the consultant reasons, will "lower sales and rent costs," allowing the poor to participate in the market.

This theory is consistent with "bottom of the pyramid" practices identified by other critical work, mostly on microfinance, telecommunications, and humanitarian design (Elyachar 2012; Roy 2010a). It is a theory of poverty that sees scarcity as a source of consumer demand rather than as a barrier to economic growth. In addition, the house itself—outfitted with off-grid micro-tech—is designed to be independent of (and therefore free from the failures of) large-scale infrastructure and administrative systems (see Collier et al. 2017). All of that off-grid technology alleviates the pressure on existing infrastructure, but it also converts infrastructure needs—water, energy, trash disposal—into new consumer markets.

Only, as I show in the following sections, the city's affordable housing needs were already being met by Kigali landowners in neighborhoods like Kiyovu cy'abakene. The consultants who worked on the house, however, did not fabricate the shortage of affordable housing that their work was supposed to solve. It was first identified in a 2003 UN-Habitat report, *Slums of the World*, then listed as an official statistic in the Kigali Conceptual Master Plan, and later published in market projection reports. In that first UN-Habitat report, tucked away on a single page is a startling statistic about the Rwandan capital: more than 80 percent of Kigali's population lives in "slums" (Moreno 2003, 25).

THE PROJECTION OF SCARCITY

Statistical projections played a key role in constructing the housing "crisis" in the OZ plan that the Batsinda house was supposed to solve. In 2003, when UN-Habitat estimated that more than 80 percent of Kigali residents lived in slums, the organization was working toward a universal definition of the concept. The issue, from UN-Habitat's perspective, was that "slum" and "informal settlement" (two words that are used interchangeably by the organization) had an impossible range of definitions that varied across space and time, making their work of measuring project impacts difficult. That year, after a series of conferences in Nairobi, the organization announced its new definition of the term. A "slum," it said, was a human settlement that lacked five measurable elements: adequate access to potable water, adequate access to sanitation, adequate access to secure land tenure, adequate access to living space, and adequate access to shelter built with durable materials (see UN-Habitat 2003, 20).

The stated purpose of the following UN-Habitat report *Slums of the World* was to count the number of "slum dwellings," or household units everywhere that met the UN's official definition. The authors of *Slums of the World* arrived at their count by compiling statistics, provided by government censuses, on their five attributes of slumness. According to the report, this method of using government numbers about unrelated issues generated its own set of limitations. Households, for example, could be counted multiple times if they reported issues with both water and shelter. To address this problem, the authors of *Slums of the World* shifted the UN's definition of "slum" from a place that lacked *all* five elements to "a group of individuals living under the same roof lacking *one or more* of the conditions" (Moreno 2003, 18; italics original). If three households reported three distinct issues with very different solutions to census takers—say one needed better access to water, another needed a way to get property documents, and another needed their trash picked up—UN-Habitat would lump them together as sharing the same problem: they all lived in "slums."

To arrive at their statistic about Kigali—that 82.9 percent of the Rwandan capital lived in "slums"—the authors of *Slums of the World* appear to have imported statistics about water, waste management, and living space from Rwanda's 2001 national census. According to their description of methods, the authors of *Slums of the World* then added up all Kigali households who experienced one or more of the five specific problems they measured. In other words, when 82.9 percent of Kigali residents are counted as living in "slums," this number represented the 82.9 percent of Kigali residents who had some sort of issue with either water, property documents, trash pickup, or living space in 2001.

While UN-Habitat's stated purpose of bringing the concept of "slum" back in global policy discourse was to identify precise metrics of slumness, its work had the opposite effect. Once the data on distinct service delivery issues were bundled into a statistic about "slums," it could be used to indicate all sorts of problems that are not reflected in the original reports. Once households that, for example, had an issue only with getting documents were reclassified as slum dwellers, they also needed more stuff—more houses, more water, more utilities, better building materials.

The 2007 Kigali Conceptual Master Plan uses the UN-Habitat numbers in a way that further unmoors the statistics from what they are meant to represent. The UN-Habitat report about Rwanda indicates only that 83 percent of Kigali residents in 2001 reported that they had some issue with *one or more* of the five conditions that UN-Habitat identifies as characterizing a slum. In the Conceptual Master Plan, however, these statistics are used as evidence of a shelter crisis, an indication that most of the city needs more houses, more water, more electricity, and more waste management: "According to available figures," writes the OZ team, citing the UN-HABITAT report, "approximately 83% of the urban population of Kigali is located within informal settlements. The implication [of the UN-Habitat report]," they go on, "is that a large proportion of the city's population is living in highly dense, substandard conditions, with poor infrastructure services that if not properly managed will result in increasingly failing environmental and urban systems" (OZ Architecture 2007, 23–24). This projected crisis of "substandard conditions" and "failing environmental and urban systems" is what the model house was tasked with converting into a market—a market whose demand could be met with the private construction of more houses.

"The need," the OZ team goes on, "for additional housing in both new and redeveloped areas of Kigali cannot be overstated. Especially for low and medium standing needs, housing models must be developed that are affordable, using appropriate technology and materials" (OZ Architecture 2007, 91). And, the plan notes, Engineers Without Borders had already built a model that was affordable and used appropriate technology and materials. "In 2006," the master plan states:[4]

> a team of dedicated citizens and engineers came together from the City of Kigali, DED, KIST, and Engineers Without Borders USA to design and construct a model house that would meet the requirements listed above. . . . The Model Home case represents the best kind of innovative outcome that can result from a concerted collaboration between public, private and civil society. (OZ Architecture 2007, 90)

In the UN reports and the Kigali Conceptual Master Plan, the language of deficit is used to construct a problem of scarcity—scarce housing, scarce infrastructures, and scarce knowledge of how to build houses. But that problem is not a barrier to growth. It is a new market—a need for expert labor and privately built sustainable houses and the technologies that keep these houses off the grid. The issue was that demand for affordable housing was already being met by the very people the plan envisioned as the target consumers for the house. And it was being met through a low-impact arrangement between the city's residents and the wetlands.

KIGALI'S IMPERFECT ARCHITECTURES

When former residents describe life in Kiyovu cy'abakene, what emerges is not the inadequacy of an "unplanned" and "informal" settlement, but the affordances of a popularly built *akajagari* (unauthorized) *katziye* (neighborhood):

> When I first arrived [in Kiyovu cy'abakene], you could *see* it was a beautiful/good/quality neighborhood [*katziye nziza*] and it was hot [*harashyushe*, vernacular for "bustling/popular"]. In the streets, you saw all people: rich people, simple people,

there were drunks, crazies [*abasazi*], prostitutes [*indaya*], there were merchants, and people who . . . you know do like normal work [*nka akazi gasanzwe*]. Really, you could find every type of person from all of society [*muri sosiyete yabantu bose ukibonamo*] . . . we were all hiding there! [laughter]. (Recorded interview with Ms. Kayitesi, Bannyahe resident, December 12, 2014)

I first met Ms. Kayitesi in Bannyahe, where she moved after Kiyovu cy'abakene was demolished. As a tenant who did not own land, she was not entitled to compensation, nor was she selected to move to Batsinda. When the house she rented was demolished, she moved to another neighborhood in the popular city and found a cheap rental there.

Ms. Kayitesi arrived in Kiyovu cy'abakene with her sister from the Rwanda-Zaire (now Congo) border in 1996 during the city's postgenocide population boom. They came to Kigali for two reasons: they wanted to get away from the intermittent war on the border, and they needed work. Both found work as mobile street traders selling produce in discount street economies.

Kiyovu cy'abakene had everything Ms. Kayitesi and her sister needed: rents were low, food was cheap, and they could walk anywhere. The rooms she and her sister rented near the Las Vegas nightclub were not exactly five-star accommodations. The sisters lived in one of four houses that faced each other around a central courtyard—one house was for their landlords and the other three were rental properties. But the sisters had electricity hookups in their rooms and access to potable water through a shared tap in the center of the courtyard. And while thrift was certainly part of Ms. Kayitesi's decision to live there, Kiyovu cy'abakene offered her something the "formal" city did not: an urban space that was tolerant of the people the "formal" city was hostile to. Far from perfect, Kiyovu cy'abakene was still a "beautiful/good/quality neighborhood" that served as a refuge ("we were all hiding there") for people who were otherwise excluded from the city center by the RPF's infamous policies against akajagari (see chapter 2). In contrast to the hyperpoliced elite enclave just across the road, Kiyovu cy'abakene offered a tolerant, cosmopolitan space, where "drunks," "crazies," sex workers, and street traders mixed with "all of society": rich people, "simple" people, and merchants.

As discussed in chapter 2, Kiyovu cy'abakene came up in the late 1960s and 1970s as new residents moved into the city and worked with lower-level officials to build their own housing stock and service delivery systems. After the genocide, Kiyovu cy'abakene—with its multiuse housing and inexpensive rental properties, all in close proximity to town—was at the center of the population boom and of economies of repair and recycling. The diversity and industry of Kiyovu and the discount economies created the "hot" aesthetic Ms. Kayitesi remembers—a neighborhood where there is no wasted space and where you can "see" a multitude of activities and people engaged in the street.

Of course, Kiyovu cy'abakene was not a perfect space where all the proletariats lived happy, joyous, and free lives. Rather, it was a popular response to state neglect and economies of scarcity. Close to the wetlands, there were ample issues with public health, such as waterborne diseases (discussed in chapter 2). The discount rental economies that thrived there served capitalism by keeping wages down, making it possible for the city's essential workers to reproduce their labor on less than a dollar a day (see Shearer 2020a). And buildings went up with a commitment to minimalism and functionality over aesthetics. Still, as Ms. Kayitesi and her former neighbors observed, it was not the inherently *un*sustainable, "informal settlement" that the master plan and so much scholarship on the neighborhood said it was. It had plenty of form. It was a tolerant space (for "all of society"), a pedestrian neighborhood where fresh food, employment, and popular transportation were available within walking distance, and it was built with near zero-carbon resources (clay, earth, sand) from the wetlands just twenty meters downhill, using low-carbon technologies (air-dried bricks).

In other words, Kiyovu cy'abakene was, to borrow from Cajatan Iheka's (2021) work on African "imperfect media," an imperfect urban space. In his work, Iheka conceptualizes imperfect media as an ecological response to what he calls, following García Espinosa, "perfect media." For Iheka, "perfect media names the typical Hollywood production; it is beautiful, promotes resource depletion, and leaves a significant ecological footprint" (2021, 226). Against what he calls the "ecologically reactionary" high-budget and high carbon–producing perfect media, Iheka offers "imperfect" popular African media productions in sound, film, and images, which are characterized by recycling, minimalism, improvisation—practices made necessary by "an overall economy of scarcity" but also reflecting a commitment to reaching a wider audience. Two of Iheka's examples of imperfect media are Nollywood and the improvised, low-budget, and lo-fi recording practices that Louise Meintjes (2017) engages with in her work on South African recording studios. Iheka argues that while these popular modes of media production reach wider audiences than their high-budget counterparts, they are often "dismissed as [low-quality] instances of amateurism or deviations from aesthetic excellence" (Iheka 2021, 226). But it is precisely because of these conditions that "Eurocentric media practices, with their characteristic high consumption models, have much to learn from their African counterparts premised on recycling and minimalist aesthetics" (Iheka 2021, 227).

Following Iheka, Ms. Kayitesi's description of Kiyovu cy'abakene as a "beautiful/good/quality" neighborhood, despite its imperfections, could certainly be read against the reactionary perfection of master-planned sustainable urbanism. Kiyovu cy'abakene's imperfect architecture—made by generations of Kigali residents who systematized and expertly constructed a neighborhood from clay—is analogous to Iheka's "imperfect media"—low-budget, low-carbon, and low-impact. Of course, not everyone in Kigali agrees with this assessment. For example, in 2014,

I asked Kigali's then director of urban planning and construction to clarify the city's definition of informality. She replied: "[W]e [the city of Kigali] have to ask, how is it built? Is it self-built? There was no professional architect, no planner, no permit? . . . Then we would classify it as informal. Also, 'informal' can indicate the lack of proper construction methods . . . or maybe it was built where it should not be. Think of it like this: informality is defined by legality. If it is informal, it is not abiding by the legal codes" (unrecorded interview, April 1, 2014, reconstructed from field notes). The land use and building codes the director referred to were written in 2012 as part of Surbana's master plan and rewritten in 2015, 2019, and 2022. The rigid, single-use zoning master plan that Surbana created (described in chapter 3), which rendered over 80 percent of the city illegal ex post facto, was published in 2013. According to the building codes and the master plan, not only was most of the city in the wrong place, as the director observed. It was also made with the wrong methods, was designed by the wrong people, and in the words of the director that day, "did not meet the basic criteria for what constitutes inhabitable."

On that day, it was explained to me that the basic criteria for what constituted "inhabitable" and "durable" materials in the building codes were architectures of concrete, steel, and glass—all materials that must be imported to Kigali at great financial and environmental cost. These perfect architectures, like perfect media, risk "being ecologically reactionary" (Iheka 2021, 227). "Durable" materials like concrete are cited as responsible for about 8 percent of the world's greenhouse gases (see Rodgers 2018).

Kiyovu cy'abakene, however imperfect, drew primarily from clay, earth, and sand as building materials. Working with other building materials such as concrete, baked bricks, and glass requires large amounts of capital and expensive machinery, restricting their use to large firms owned by the ruling party. But as discussed in chapter 2, earth for building is everywhere in Kigali, and there were large sand deposits across the wetlands in Kimicanga. (Clay is also earthquake and fire resistant.)

As discussed in chapter 2, while building typologies were diverse, much of Kiyovu cy'abakene was built with two prevailing modes of construction that use clay from the wetlands with a variety of other materials: wattle and daub (*ibiti na ibyumba*) and air-dried bricks (*amatafari ya rukarakara*) (see figure 9). In both cases, construction is a labor-intensive, low-carbon process that uses little (if anything) in the way of machinery or fossil fuels. And, like 60–90 percent of Kigali's built environment (depending on whom you ask), most of the structures in Kiyovu cy'abakene were sealed with *karabasasu*, a mixture of coarse sand, lime, and/or cement that seals the walls, protecting them from rain and insects while slowing decay.

The easy availability and affordability of brick construction—air-dried bricks go for five to ten cents each, and working with wattle and daub is even

FIGURE 9. Amatafari ya rukarakara waiting to be used in a construction project. Photo by Samuel Shearer, 2021.

cheaper—supports a system of modular development. Air-dried, compressed clay does have a short shelf life, but this is a plus in a city where the built environment routinely shifts to accommodate the city's fickle economies. In contrast to more "durable" materials like cement or burnt bricks, the clay structures in neighborhoods like Kiyovu cy'abakene allow for an architecture of anticipation. Everything in the popular city goes up with the intention that it will be adjusted, subtracted, and repaired to meet the evolving needs of its inhabitants; the impermanence of clay enables this continual adaptation of the city's popular spaces and the economies they support.

The results are not "informal" settlements but an urban form that allows every inch of space to be employed in the production, circulation, and reinvestment of value into the neighborhood. A landowner might start with a single-room house and then add an annex that can be rented. Surplus rents will then be reinvested into more built space that can provide an alternative source of income—a barbershop, for example, or a small bar—that will offset the risks of relying on rents. As the landowner reinvests resources into the building, walls will be taken down to make room for additions, and more walls will go up to create new spaces. This kind of flexible built environment is simply not possible with more "durable" building materials.

Kigali's imperfect, low-carbon, low-impact architecture sustained Kiyovu cy'abakene for decades. Air-dried bricks kept rents down by keeping construction costs low, which also kept the affordable rental housing market in the hands of Rwandan landholders.

Despite its issues, Kiyovu cy'abakene did meet many of the requirements of textbook "sustainability." The neighborhood was walkable; centrally located; fully serviced with electricity and water, its own fresh food market, and schools; and built primarily with low-carbon construction technologies. Landlords lived on-site and invested their surplus capital into repairing and improving the built environment around them (as opposed to off-site slumlords who extract capital from communities). And it was that reality—of a large affordable housing stock in the center of town—that threatened the crisis of housing scarcity the Batsinda house was supposed to solve.

Evidence that the Batsinda house was designed to address an issue of housing scarcity is clear in the Kigali Conceptual Master Plan and the consultants' writing on the house discussed above. But it was also reinforced by housing market projections that are used to advertise Kigali as an investment destination. After the projections of housing scarcity in the OZ master plan, the City of Kigali hired a French business consultancy to write a ten-year housing market projection report. According to the math in the report, the firm projected a ten-year deficit of 344,068 housing units that it counted as effective demand for new housing, of which over 54 percent would be for affordable housing (Planet Consortium 2012, 25–26). Only that deficit is not a deficit at all. Over 30 percent of the projected deficit is extrapolated from what the report counts as 108,803 *current* units, or housing "backlog" (Planet Consortium 2012, 30). Destroying the backlog houses, the authors of the report reason, is key to kick-starting Kigali's international housing market. This is because demolishing over 100,000 houses owned by individual Rwandans will free up an estimated 727.62 hectares of land for "redevelopment" by private real estate firms (see Planet Consortium 2012, 18). As people are dispossessed of their property and scarcity is generated through demolition, they also become consumer demand for newly built affordable housing (see also Uwayezu and de Vries 2020a, 4).

A later update to the report written by another outside consultancy warns that this backlog statistic only represents the number of houses that show up on government reports that do not align with the city's strict building regulations. They do not know the quality of backlog houses as they are going off census figures. All they know is that the majority of the backlog are counted because of the materials they are made from: clay (see Bower et al. 2019, 52). "This observation," the report goes on, "makes clear the scale of the impact of the zoning codes or building regulations that disallow houses with improved, covered, earth bricks" (Bower et al. 2019, 52). But while the report correctly worries that building codes are distorting the numbers of "backlog houses," the city's housing market reports also reveal something much more profound: most of the city's housing deficit is not a deficit.

It is the number of homes that do exist but must be demolished to generate effective demand for new housing markets.

In other words, the actual threat of "informal" housing in the urban core was not to the environment or urban systems, nor was it because ruling party authorities did not like how it looked. Kiyovu cy'abakene—an imperfect but nevertheless affordable and low-carbon solution to issues of housing in Kigali—posed a threat to the promise of Kigali as a new market for affordable housing and "green" service delivery systems.

THE CONSTRUCTION OF SCARCITY

"This place," Ms. Munezero said with a laugh, "is fake [*aha hantu ni fake*, using the English word *fake*]" (recorded interview, July 24, 2015). When I spoke to her in her "fake" Batsinda house in July 2015, Ms. Munezero was one of the last remaining residents in the Batsinda Housing Estate. Ms. Munezero's family belonged to the first group of Kiyovu cy'abakene evictees, who, unlike Ms. Ingabire and Ms. Muhizi (see chapter 3), quietly accepted being relocated to the Batsinda Housing Estate. "If you don't cause trouble," she explained matter-of-factly when I asked her why she hadn't fought the move like some of her neighbors, "they [city officials] let you keep your furniture."

Before the demolitions, Ms. Munezero lived in Kiyovu cy'abakene with her parents in a four-bedroom home on a plot that also had two smaller houses with shared access through a courtyard to a communal kitchen and bathroom. Like everyone who was moved to Batsinda, they were homeowners. They rented their smaller houses out. Her mother also sold vegetables in Kiyovu market, and her father worked as a day laborer in construction. When the family was moved to Batsinda, they lost their rental properties and access to the market, not to mention the advantages of being in the city center.

There is significant disagreement between Ms. Munezero, who had to figure out how to use the Batsinda house, and the consultants who designed the prototype over whether the house was an "improvement." On its own, the Batsinda house, which Ms. Munezero called fake and the designers called a model of sustainability, could certainly be considered an ingenious solution to the shelter crisis as it was presented in city plans and housing market reports. It was a two-bedroom, single-family home made with all local materials and outfitted with technologies that address all five issues that UN-Habitat identified as problems shared by "slum-dwellers" throughout the world.

For the designers, their housing prototype also made significant improvements to the visual aesthetics of what they called "typical informal housing" in Kigali and "achieved a very modern look . . . compared to other local construction" (Ilberg 2008, 6). As evidence of "local construction," they present an image of a few structures that appear to be in various stages of construction, captioned "typical

FIGURE 10. A single plot in Cyahafi, a neighborhood in the urban core. Note the X next to the door on the bottom right, indicating that this is a "backlog" house set to be destroyed. Also note how this house, unlike the one in figure 11, is clearly connected to the grid, just as Kiyovu cy'abakene was. Photo by Samuel Shearer, 2022.

traditional housing in Kigali" (for this image, see Ilberg 2008, 6). The image shows a structure with a sagging roof and walls that are either still being worked on or have not been maintained. But the consultants' image of the house is not typical of housing in Kigali. In fact, the house they present as typical appears to be in an area that is quite rural—and there is no evidence that it is, like Kiyovu cy'abakene, connected to the city's hydraulic and electricity infrastructure.

Figure 10 is a better representation of the houses Ms. Munezero and her neighbors were forced to leave: a multiunit property with living, rental, work, and retail space on one plot, located in Cyahafi, a neighborhood near Kiyovu cy'abakene that shares Kiyovu's postcolonial history and socioeconomic makeup. Like Kiyovu cy'abakene, this neighborhood is owned and operated by the people who live there. Located just above the wetlands on the other side of Nyarugenge Hill from Kiyovu, it also has the same geological conditions. It is serviced by the same water and electricity base, and like Kiyovu, it has also been slated to be demolished by the master plan.

Like many of the buildings in Kiyovu cy'abakene, all the structures in figure 10 are held up by load-bearing walls built with amatafari ya rukarakara. Each wall is sealed with karabasasu (a mixture of sand, paint, and lime) to protect the walls from rain and insects, giving the house its beige coloring. Unlike the sustainable prototype house, this is not a single-family arrangement. The main building is a two-unit house with separate entrances. To the left of the main building is an annex (with a satellite dish on its roof) that is a rental property. Two more annexes have been built behind that one; these are open, one-room structures that can serve as rental properties. Building on this plot probably began with a single one-room structure with an earth foundation that was later upgraded to concrete. Over time, the owners made incremental adjustments, using the malleability of clay to their advantage. Notice, for example, the stand-alone wall on the far right, where a wattle-and-daub extension has been started; over time, it will become the next

FIGURE 11. An image of the prototype Batsinda house as presented in the 2007 Kigali Conceptual Master Plan (p. 90) and the Planet Consortium / City of Kigali report "Housing Market Demand, Housing Finance, and Housing Preferences for the City of Kigali" (p. 18). Neither document credits an author for this image. Note the off-grid infrastructure to the right of the house. The black tank is a rainwater catcher that sits on top of a biogas converter. Source: City of Kigali / Planet Consortium.

structure on the plot. The faded *X* just to the left of the rightmost door on the main building indicates that these buildings belong to the "backlog" listed in housing market studies and slated for demolition.

According to Ms. Munezero, the Batsinda house was "fake" because it had none of the adaptive features of the building in figure 10. Unlike the multifamily, multistructure construction in Kiyovu cy'abakene and neighborhoods like it, the houses in Batsinda had no spaces to rent, sell from, or put to other productive uses. They were houses for people who—unlike Ms. Ingabire, Ms. Muhizi, and Ms. Munezero—lived in nuclear families in one place and went to work in another (see figure 11). The house was a closed, finished design with discrete spaces set up solely for the unpaid labor of consumption and social reproduction: cooking, sleeping, bathing, and childcare. Ms. Munezero's rejection of the Batsinda house as "fake" is certainly an example of what Claudia Gastrow (2017a, 2024) calls "aesthetic dissent," or the refusal to be interpellated by a designed rendition of "progress." But the Batsinda houses were also fake because they were missing key elements that city residents expected to be in a house.

When I asked Ms. Munezero about those first years in Batsinda, she said what everyone did: there was no water or electricity for the first year, and no way of getting to the city. "We had to fetch water from the wetlands, like *abaturage* [peasants]! [laughter] It was bad . . . we had to use candles for light . . . we couldn't go to the city, and there was *nothing* to do here [*nta ibikororwa birahari*]."

"What about your rainwater catcher?" I asked, thinking about my own daily trips to fill jerry cans of water where I lived in Nyamirambo. "You mean that tank in the back?" she asked. "It's there [*irahari*]." She sighed. "It broke in the first month. We can't fix it. There are no spare parts."

For Ms. Munezero, the absence of water and electricity was clearly a failure, oversights or corners that were cut to keep costs down. The absence of service-delivery systems, however, was not a failure from the perspective of sustainable design. It was a design feature of the house, advertised as part of what made it green. "Critical to the project," the consultants write, "was 'off-the-grid' style infrastructure" (Ilberg 2008, 6) so the house would not tax the city's water and electricity grids, saving money for the municipality. The Batsinda house was designed to get its electricity from a solar photovoltaic energy system manufactured and sold by Great Lakes Energy, a US-owned company that works out of Rwanda (see Ilberg 2008, n. 18). The house's water comes from a rainwater catcher. And the house's fuel comes from a biogas converter, a privately manufactured technology that is advertised to take care of both sanitation and cooking fuel at the same time.

All of this off-grid technology with near zero carbon footprint also represents new markets for green commodities. Every off-grid house that goes up also generates consumer demand for solar panels, biogas converters, and rainwater catchers—and the spare parts to maintain them. So, while the Batsinda house did not generate value for the people who lived there, it did promise to generate consumer demand for private firms selling solar panels, rainwater catchers, and other green tech to residents disconnected from Kigali's grid.

But the Batsinda houses were also similar to the houses in Kiyovu cy'abakene in one important way: the building technology at its core, which is now presented in engineering textbooks as Engineers Without Borders' innovation (see Striebig, Ogundipe, and Papadakis 2015, 21–22). It is a building technique that "reduced by a factor of *five* the established cost estimate" for building a comparable house from cement, steel, and glass (Ilberg 2008, 5; italics original). This technology leaves little carbon footprint; it is renewable; it is affordable.

It is an air-dried brick made from local clay.

THE DISPOSSESSION OF KNOWLEDGE

Tlotlo Tsamaase's (2020) Africanfuturist short horror story "Behind Our Irises" takes place in Gaborone, Botswana. It is told in the first person by a narrator who lands a job with a European consulting firm that has offices in the city. After a probationary period with the firm, the narrator is offered an "upgrade" with benefits and a better salary. The upgrade involves Tsamaase's narrator taking a pill that installs software in their head. The pill also stops the narrator from screaming when they notice that the firm has drilled holes into the back of their colleagues' necks. The holes are used by the consulting firm to extract collective technologies, designs, and creative innovations from the brains of their employees. This knowledge is then "aggregated into its [the consulting firm's] network to create 100% authentic indigenous products, used for concepts in fashion shows, architectural designs to win local tenders" (Tsamaase 2020, 50).

Tsamaase's short story is about many issues: workplace sexual assault, the continuation of racism in industries that claim to disavow it, and the plunder of African knowledges and technologies by corporations. In the story, the upgrade pill facilitates the transfer of creative ideas from the firm's African employees to outside consultants while silencing those who are dispossessed of that knowledge.

Unlike the firm in Tsamaase's short story, the consultants who worked on the housing upgrade did not set out to extract knowledge of how to make a house with air-dried bricks in Kigali only to sell bad copies to people who already had their own homes. Their stated intentions were to use the house to resolve a debate over sustainability among different authorities in the city of Kigali. They aimed to defend amatafari ya rukarakara against its detractors who also worked for the government. Their goal was "to legalize earth as a building material" (Ilberg 2013, 155).

The consultants agreed that amatafari ya rukarakara was "affordable and sustainable infrastructure alternatives [to building with other materials]" (Ilberg 2013, 155). In their view, air-dried brick was more sustainable than other "durable materials" like burned brick, "which causes deforestation and significant carbon and particulate emission when locally produced. In fact," they go on, "the burned bricks are often no better than traditional adobe when tested in compression or durability, yet the environmental impact from their manufacture is enormous" (Ilberg 2013, 150). In their words, the Batsinda house was designed to "change common misperceptions" (Ilberg 2013, 155). The house that did not work for Ms. Munezero, because it was single family, enclosed, and had no spaces to rent, was designed to make clay more aesthetically appealing to city officials. As they later wrote, "[u]nfortunately and unpredicted, the availability of the housing concept may have accelerated evictions from other neighborhoods and involuntary resettlement" (Ilberg 2013, 155).

Even though it was not their intention, the consultants' "upgraded" housing model, like the upgrade pill in Tsamaase's short story, facilitated the dispossession not just of land but also the knowledge of how to build a house with local earth. This is because the Batsinda house was designed to do much more than "change common misperceptions"—to also generate market growth in a way that transferred near zero-carbon technology that was used to build Kiyovu cy'abakene (amatafari ya rukarakara) to private firms. This required more than involuntary resettlement: a process of cheapening Kiyovu cy'abakene and the generations of knowledge of how to work with clay through the language of "informality." Homes in Kiyovu cy'abakene that were built with the same materials as the "sustainable" house (sealed, air-dried earth bricks) were described as "informal" and therefore *unsustainable*, erasing the origins and the value of sustainable technology (air-dried brick) from Kiyovu. Even though there is ample documentation in the consultants' writing on the house, the demand for reparations, and how former Kiyovu residents understood their displacement, the dispossession of knowledge is persistently overlooked in accounts of the demolition of Kiyovu cy'abakene. It is, to borrow from Michel-Rolph Trouillot (1995), "unthinkable" within the language of deficit that is used to describe Kiyovu cy'abakene that its residents, as urban Africans, already had the capacity to build ecologically significant, low-cost, durable housing. What Tsamaase's work makes thinkable is a process that dispossesses people not just of their homes but of the value of the technology that was used to make those homes.

THE PRODUCTION OF SCARCITY

In 2006 the City of Kigali partnered with the national pension fund and used the housing prototype to develop a pilot sustainable housing program. The pension fund fronted the costs to build 250 homes in Batsinda, on what was then the edge of Kigali. These homes were then put on the market at RWF 3.5 million (about a 60 percent markup from the construction costs).

City authorities and planners directed considerable promotional hype on behalf of the Batsinda Housing Estate. Planners who worked for the City of Kigali gave foreign journalists guided tours of Batsinda. One US journalist reported back home about an "environmentally progressive haven" where "rainwater collection systems grace every rooftop, providing water and keeping erosion in check. Biogas digesters behind communal latrines miraculously transform human waste into fuel for cooking" (Marshall 2008, 26). UN-Habitat also hyped the project internationally, making it the centerpiece of Kigali's 2008 Scroll of Honor Award (Kimenyi 2008). There was no talk of demolition. As one planner at the City of Kigali told a US journalist, "There is already a real shortage of low-cost housing in Kigali that is planned and decent" (Marshall 2008).

But the empty Batsinda Housing Estate stood as proof that the "crisis" of affordable housing in Kigali existed only in UN-Habitat documents, master plans, and housing market reports. Kigali's affordable housing needs were being met elsewhere, in neighborhoods like Kiyovu cy'abakene, where new arrivals to the city like Ms. Kayitesi and her sister found centrally located, low-cost rentals owned by landlords like Ms. Munezero and Ms. Ingabire.

The City of Kigali turned its attention to Kiyovu cy'abakene for two reasons: first, as the mayor of Kigali explained to the press before the demolitions, the plan was to demolish everything between the original core and the wetlands to make real estate available for the new city center (Rwanda News Agency, February 5, 2008). Second, because the Batsinda houses were built with technology that Kigali landowners already controlled (amatafari ya rukarakara), the only way to generate effective demand for the Batsinda houses was to clear the "backlog" houses that were outcompeting Batsinda with housing that was more affordable (as implied later in the housing market reports).

Importantly, no one from Kiyovu cy'abakene opposed being moved. After all, they knew how to build and rebuild their neighborhood. They had done it before. The problem, Mr. Musubimana, a former Kiyovu cy'abakene resident who had been moved to Batsinda, said, was *how* they were moved: "*batwimutse nabi* [they moved us badly]." The residents wanted to negotiate fair financial compensation for their land and houses, which required the neighborhood, its architectures, and its infrastructures to be recognized as valuable. As Ms. Mutiyimana, a third-generation landowner from Kiyovu, told me, "They [city authorities] told us they wanted to build expensive buildings [*amazu y'agaciro*, valuable houses] in Kiyovu . . . I don't know, but maybe like what you see in the master plan." She had no problem with that. "If we saw enough money," she said, "we would have left—wouldn't have mattered [*ntacyo bitwaye*]" (recorded interview, July 26, 2015). The issue for her was that Kiyovu cy'abakene had become devalued through the discourse of informality.

The official record states that Kiyovu landholders were offered compensation of RWF 3,563 per square meter (about $6.65 per square meter in 2008) (see Mugisha 2015, 8), although many people I interviewed said they were offered far less, and some were offered no compensation at all. The plan was to transfer the small plots that were owned by large numbers of Kiyovu families into the hands of a single holder—the national pension fund, which would then put the land on the international real estate market for $200 per square meter, a massive markup (see Bishumba 2010). The pension fund would also get 126 families to occupy the empty houses in Batsinda. Those moved to Batsinda would have their compensation subtracted from the RWF 3.5 million cost of the house and would then be responsible for paying off the rest via a mortgage provided by the pension fund. Their tenants would be on their own.

That was the deal a group of landowners refused.

"We knew we had to move," Mr. Hakizimana, a former Kiyovu landowner, said, "but . . . we thought we would get enough compensation to buy land somewhere else" (recorded interview, May 8, 2015).

"People refused [*banze*] their compensation," his partner, Ms. Hakizimana, interjected, "so they brought in the Caterpillars and demolished [Kiyovu cy'abakene]."

Mr. Hakizimana then backtracked to the compensation scheme. He laughed as he told the story:

> We [residents] had a lot of meetings [with the government] . . . we had to sign a lot of papers . . . but kid [*sha*, vernacular], they worked us over [*baduhenze*]! . . . For example, we had two houses on one plot. You know how much they offered me? 300,030 [about $500 in 2008]. You can't buy anything with that! [laughter] Then Caisse Sociale [the pension fund] told me, they said, "Your compensation is not enough. Go to Batsinda and we will loan you the 3.5 million francs!"

Mr. Hakizimana was now struggling to speak through his laughter: "You understand [*wumvise neza*]?" he asked me. He never saw any of the promised compensation. Instead, the pension fund applied it to a mortgage that he did not want. "They demolished *my* houses, took *my* property, and instead of compensating *me*, I owed *them* 3 million francs!" [more laughter] (recorded interview, May 8, 2015; see also Mutara, December 3, 2008, for Rwandan media sources that confirm these accounts). Importantly, for Mr. Hakizimana, Ms. Mutiyimana, and most of their former neighbors, being displaced was not what they saw as the *akarenganye* (injustice). It was being stripped of the value they had built up in their homes, shops, and rental properties, and being coerced to consume "humanitarian" sustainable homes they did not want.

Everyone—Kiyovu landowners and press accounts—agrees about what happened next. In early July 2008, a group of Kiyovu landowners filed an appeal with the Office of the Ombudsman. The ombudsman sided with the landowners: they were being undercompensated for their land. The Office of the Ombudsman halted the demolition, placing an injunction on the move until the group of Kiyovu landowners who were holding out could hire independent appraisers and get an accurate value for their properties (see Wakhungu et al. 2010; also see Gakire 2008 and Kalisa 2008 for press reports that were written shortly after the demolitions). Its plan to use economic force to dismantle Kiyovu stymied, the City of Kigali turned to extraeconomic violence—demolition.

In an interview at Ms. Muhizi's place, with her, Ms. Ingabire, and Ms. Ingabire's husband, we went over the events leading up to and after the demolition of Kiyovu cy'abakene.

Ms. Ingabire, who remembered the goat being crushed (see chapter 3), jumped in: "People ran out to the tarmac road that goes to Kacyiru." She turned to Ms. Muhizi: "What's his name? The [cell] executive?"[5]

"E—," Ms. Muhizi said.

"Him," Ms. Ingabire continued, turning back to me. "He was yelling [at the day laborers], '*Musenye! Musenye!* [Demolish! Demolish!]'" Ms. Muhizi nodded in agreement.

According to Ms. Ingabire, the workers froze, unsure of what to do. Behind them was the cell executive screaming orders. In front of them, between the workers and the houses, rocks in hand, stood about a dozen Kiyovu landowners. "That's when Kirabo [the then mayor of Kigali] arrived," said Ms. Ingabire.

It was the mayor, she said, who called in the Rwanda National Police. "They must have sent in all the police trucks in the country!" she said. "All of them [*zose*]!" Ms. Muhizi agreed.

Residents who protested were handcuffed and forced to sit on the road and watch the workers finish the job the bulldozers began. Then, Ms. Muhizi said, "They ordered us, 'You get in that bus there [*mukwinjera mumodoka hariya*]!' We got on the bus and then *shoosh!* . . . they rushed us out [of Kiyovu]!"

They were taken to Batsinda—a place far away from Kigali's edge that had no electricity, no running water, and no transportation options back to the city. "We were beaten/defeated [*twarakubiswe*]!"[6] Ms. Ingabire laughed. Ms. Muhizi nodded.

By the end of the day, scores of affordable homes and workplaces on 362 plots, a market, and a nightclub were destroyed, and Kiyovu residents who once supplied affordable housing close to the city center were coerced into becoming consumers of "fake" houses they did not want (see also Kalisa, 2008).

Where Kiyovu cy'abakene once stood mostly remains, as many former residents point out, *amatongo gusa* (only ruins) with the exception of a few high-rise hotels and government buildings built with concrete, glass, and a large carbon footprint. The emptiness that former residents see there reinforces the notion that the destruction of neighborhoods in Kigali is not only about the accumulation of land by dispossession but also about "the production of scarcity" (Barnes 2014). In other words, the demolition was not only, as the consultants who designed the house later wrote, used by an unnamed agent "to justify expropriation for private land development" (Ilberg 2008, 8). The demolition was also used to justify the model house.

Kiyovu residents did not, however, simply accept the move. They pushed back with unauthorized repairs.

REPAIR

In February 2015 Mr. Gasore, Ms. Ingabire's husband, brought my attention to the demand for reparations that they made to the state, quoted at the beginning of this chapter. I had to ask him: "What did you hope would happen [after the letter]?"

"We hoped," he said, "they [Parliament] will get us the correct compensation for our losses from the move" (recorded interview, February 20, 2015). They wanted to be compensated for the correct value of what was destroyed in the move.

With their rental properties and businesses destroyed, the 126 families who were moved to Batsinda had no source of income. Because their compensation had been applied to their mortgages, they had no capital. Disconnected from the grid, they had no access to services. And the only way to get into the city, until 2012, was a three-hour journey on foot or an expensive moto taxi ride. They did not have the deeds to their houses in Batsinda, so they could not sell them.

"Did you think," I asked Mr. Gasore, "the state [*reta*, from the French *l'état*] would help you?"

"Our problem," Mr. Gasore corrected me, "is not with the state. It is with Caisse Sociale [the pension fund] and the City of Kigali."

The only problem they had with the national state, Mr. Gasore explained, was that the state was *un*involved in the entire process. In Mr. Gasore's account, multiple agencies that are often lumped under a singular "state" were working with and

FIGURE 12. Where the Batsinda Housing Estate once stood. Photo by Samuel Shearer, 2023.

against each other leading up to the demolition: foreign consultants, city authorities, and later the pension fund. The demolition was also driven by ideas circulating in reports produced in UN-Habitat offices. In his view, the job of the state was to repair the damage done by the demolition, which required recognizing the actual value of Kiyovu cy'abakene.

"If someone has a problem," Mr. Gasore explained to me, "they shouldn't keep quiet. It is the state that is supposed to find a solution to their problem."

On December 2, 2010, two years after the demolition and shortly after Batsinda was retrofitted with electricity, the pension fund sent an administrator to meet with Batsinda residents, not to restore justice but to get the loans paid. "No one had paid [their loans]," Ms. Ingabire laughed as she explained it during our interview in 2015. Many couldn't pay, having lost work and incomes from rentals. Others refused, going on mortgage strike.

In 2012, facing a deficit that was close to what it started with when the homes were empty, the pension fund gave in and allowed Batsinda residents to sell their homes, pay off their mortgages, and keep whatever profits were left over. Most of the former Kiyovu residents found local speculators and members of the Rwandan diaspora who were willing to buy their plots for more than the RWF 3.5 million mortgages. The new buyers demolished the sustainable houses and built brightly colored bungalows behind security gates (see figure 12).

Also that year, Kimicanga, a neighborhood across the wetlands from Kiyovu cy'abakene, was demolished (see chapters 1 and 3), sending thousands of displaced landowners in search of a new place to live.

FIGURE 13. "Dushenye" (a moniker, "[they] demolished us," from the official name, Dusheni), just outside of the Batsinda Estate in Dusheni Kagugu. Photo by Samuel Shearer, 2023.

The two groups converged to create a new neighborhood. With their profits from the sale, many former Kiyovu cy'abakene residents moved across the road from the Batsinda Housing Estate; evictees from Kimicanga and Gacuriro also moved to the same area. They worked with the same institutions that the pension fund and the City of Kigali had used to demolish Kiyovu cy'abakene: local-level (*umudugudu*) authorities supervised transactions as the newcomers bought plots of land from *bakavukire* (people born there) across the tarmac road from Batsinda. Earth continued to be banned as a building material to, ironically, protect the sustainable housing market from popular competition.[7] So the new residents hired *abafundi*, who worked at night, using contraband bricks made from clay harvested from the wetlands below and sand from a deposit near Batsinda. Former residents of Kiyovu cy'abakene, Kimicanga, and Gacuriro paid community security units and local authorities, who belong to the same institutions that were mobilized by the city to tear down their neighborhoods in the urban core, to look the other way as they built their new homes and businesses. They constructed dozens of houses and rental properties. They repaired the car-free, *akajagari* city run on economies of recycling and imperfect architectures of anticipation that the Batsinda Housing Estate had broken. They also installed ruling party institutions and voted in *abayobozi* (local leaders). The area where this neighborhood is built is officially called Dusheni. Carrying on the tradition of hijacking official nomenclature, former residents of Kiyovu cy'abakene and Kimicanga like Ms. Ingabire who were evicted from the urban core call it Dushenye ("they demolished us") (see figure 13).

CONCLUSION

It would be tempting to conclude that the Batsinda Housing Estate is a "failure" of what proponents call sustainability and critics call green capitalism. The Batsinda house never achieved the synergy of market growth in an environmentally conscious way. But today the house is showcased as a success story in venues far beyond Kigali. It features in at least one US engineering textbook published seven years after the destruction of Kiyovu cy'abakene as an example of how the best practices of sustainable engineering can convert scarcity into environmentally progressive solutions, with no reference to the violence and destruction that underwrote the project, and with no citation of the Rwandan origins of the technologies and knowledge that were used to build the house (see Striebig, Ogundipe, and Papadakis 2015, 21–22). And in Kigali today, outside firms continue to build incentivized "affordable" private housing projects to meet the city's projected housing deficit. The results are always the same: buildings stand empty until city authorities produce housing scarcity by destroying the existing low-carbon, clay housing stock, coercing people who already have homes to consume houses they do not want (see Uwayezu and de Vries 2020b; Nikuze, Sliuzas, and Flacke 2020; chap. 5 in this book). This suggests, as this chapter has argued, that discourses about African cities in crisis with green commodities as the solution are more than just bad ideas. They are central to new modes of dispossession and capital accumulation. Projects like the Batsinda Housing Estate are built around converting discourses *about* crises into profitable problems to solve that then must be made real while converting already existing low-carbon, low-impact, popular technologies like amatafari ya rukarakara into cheap capital through dispossession.

But it is not all destruction and displacement. The specific building technologies of repair that went into making Dushenye are more complex and involve many state actors who can be enrolled to work *against* the dispossessing processes of green capitalism. Making the neighborhood included much more planning and anticipation than I have outlined above. The details of how Kigali residents repair and maintain akajagari spaces by disaggregating what is already there and reconfiguring it is the subject of the next chapter. Kigali residents continued to design and build neighborhoods that incorporate architectures of anticipation, building spaces that are designed to be torn down and moved faster than city authorities can coerce people into housing that no one wants. And one of those neighborhoods, built after the demolitions began, showed how sustainable urbanism can malfunction while posing a problem for a multimillion-dollar corporate investment project: Bannyahe.

5

Repair

Punk Urbanism

Until it was demolished in 2022, all roads to Bannyahe led through Nyarutarama and Kibagabaga, two elite enclaves built in the early 2000s.[1] Bannyahe's neighbors were mostly expatriates and A-list Rwandans who hid behind spiked security gates reinforced with CCTV and armed private security guards. The only street life in Nyarutarama and Kibagabaga were pedestrian commuters moving to and from Bannyahe.

Arrival in Bannyahe brought an unmistakable shift in the atmosphere.

In Bannyahe, the main road was crowded, used for work, leisure, and hustling. *Abazunguzayi*, street traders (mostly women), pushed through the pedestrian and motorcycle traffic with baskets of produce on their heads or at their feet while men and women sold secondhand clothing on both sides of the thoroughfare. The road was lined with about a dozen "studios," small wooden kiosks that offered a variety of services from phone charging to downloading music and films from the internet onto flash drives. Behind the studios were several barbershops, hair salons, and small shops selling household items. Interspersed between the barbershops and street studios were small cafés.

If you kept walking down Bannyahe's main road, toward the cell office, you would find the private health clinic, located across from the largest bar and dance club in the area, Kwa Marico.[2] On weekends, Marico hosted all-night tear-the-roof-off-and-burn-it-down dance parties that, because his place was far away from tourists, never got shut down for busting Kigali's eight-decibel noise limit. Continue downhill, and you would eventually end up at the neighborhood's soccer pitch, the main market, and the wetlands.

Turn left anywhere off the main road as you walked downhill, and you would find yourself in a maze of houses and pedestrian-only corridors, some only a few feet wide, threading between buildings. In this area, Mukidelenka ("for delinquents"),[3] the street economies that require a bit more discretion thrived: bars that served homemade beer, unlicensed Pentecostal churches. From Mukidelenka, you could squeeze through Bannyahe's narrow footpaths, which doubled as stormwater drains, until you arrived at the middle-class interior. In this part of the neighborhood, Bannyahe's landlords, small-business owners, and a close group of active and retired soldiers in the Rwanda Defence Force built low-rise houses for themselves and as rental properties, with lush courtyards and gardens. The residential buildings extended downhill, toward the wetlands, with the built environment becoming more crowded and rents lower farther downhill. Kigali's essential workers—teachers, motorcycle taxi drivers, construction workers, musicians, and artists—lived in these less expensive areas, on plots with one or two houses. Just on the edge of the wetlands was *mumanegeka* (the high-risk zone)—a thin strip of land where flooding was frequent, residents were continually exposed to waterborne diseases, and land was dirt cheap.

Bannyahe, the neighborhood, did not end where its built environment did. The wetlands just below the neighborhood were also a source of building materials, work, and sanitation infrastructure. Abayobozi (community leaders) determined the location for shallow pits that abafundi used to win the clay that was essential to putting the neighborhood together and maintaining its form (see chapter 2). These pits were interspersed with small and medium-sized fields that grew sorghum and sugarcane. Despite its name, and despite the absence of a citywide central sewage system, Bannyahe did have access to a waste management system that many other neighborhoods built after the plan did not: a three-pond filtration system that was set up near the wetlands to convert effluent into fertilizer for the agriculture just below it (see Rwanda Environment Management Authority 2013, 90) (figure 14).[4] In addition to these infrastructures, Bannyahe was also part of citywide economies of disaggregation and discounting that provided food and clothing to most of the city.

Bannyahe certainly had its share of issues—the poorer you were, the closer you lived to the wetlands and the more you were exposed to environmental hazards of disease and flooding. And the sewage filtration ponds were not used by everyone in the neighborhood. They needed maintenance, and their functioning depended on low-paid workers whose lives were cheapened as they emptied the neighborhood's thousands of latrines while being exposed to the environmental and health hazards of working with waste.

But anyone who spent time in Bannyahe might pause at descriptions of the neighborhood as a site of "bare life" and "political abandonment" (see Hudani 2024, 134–35)—especially since Bannyahe residents, in the court cases that followed efforts to demolish the neighborhood, insisted on both the presence of governing institutions and the value of their neighborhood as more than bare. There were many government services that were built and supported by the people who lived there, and much administration in the neighborhood. Landholders in Bannyahe built water and electricity infrastructure that they then turned over to the municipal utility company. They installed ruling party security and surveillance mechanisms following RPF blueprints. Daily operations were run by an elected committee of abayobozi who reported up to cell and sector officials. Young men were hired as *abanyerondo*, a government program of community policing that landowners paid for through monthly *umutekano* (security fees). Three volunteer community health workers, trained by the Rwanda Ministry of Health, lived in the neighborhood and distributed free essential, lifesaving services—mosquito nets, malaria medication, antibiotics for typhoid—that also mitigated the hazards of living near the wetlands.[5] Yet despite the presence of governing institutions, life-extending services, and multiple layers of security, everyone—people who lived there, the press, city authorities—agreed on one issue when it came to Bannyahe: Bannyahe was akajagari, an unauthorized neighborhood where economies and building practices that had been made illicit by efforts to implement plans were tolerated.

Bannyahe followed many of the conventions of city building established in Kiyovu cy'abakene and Kimicanga. But it was also different from these neighborhoods in one significant way. Unlike Kiyovu cy'abakene and Kimicanga, which were built during the early independence period and then repaired after the genocide, Bannyahe was an act of unauthorized repair from the destructive processes of sustainable urbanism. It was built in the first decades of the 2000s primarily (although not exclusively) by people who had been evicted from the demolitions in the urban core. They did it by rendering inoperable the very mechanisms designed to prevent them from building more akajagari from the ruins of master plan destruction.

Chapter 4 followed the destruction of Kiyovu cy'abakene, a neighborhood that was built, maintained, and repaired by three generations of Kigali residents. Following the actual processes that led up to the destruction of Kiyovu cy'abakene, I showed that the destruction of neighborhoods like Kiyovu cy'abakene should be understood as more than a "state" agenda to eliminate "informal settlements"; it was about the manufacture of effective consumer demand for green commodities. This chapter picks up where chapter 4 left off. It is about the vital labor of unauthorized repair and maintenance after sustainability-induced catastrophes.

I was first invited into Bannyahe in 2013 by a longtime friend who lived in the neighborhood and who vouched for my research. For the next two years, I traveled to Bannyahe several days a week from where I lived in Nyamirambo by bus. Over those two and a half years, I conducted 113 recorded interviews and around 50 unrecorded interviews with Bannyahe residents, ranging from abafundi (construction workers) to abazunguzayi (illicit mobile street traders), security agents, landlords, teachers, local leaders, and sanitation workers, and I developed life histories of three families of *bakavukire* (people born in the area on inherited land). I also participated in six umuganda (monthly compulsory community service days) in Bannyahe, during which we maintained footpaths, drainage infrastructure, and the main road. I was there because I wanted to know where people went after demolition and how they repaired their neighborhoods. After my uninterrupted stay in Kigali ended in 2015, I returned and visited friends every year until 2022, when the neighborhood was demolished. Because this chapter challenges other depictions of Bannyahe as well as scholarship that dismisses popular politics in Kigali, I also draw on many secondary sources—YouTube videos, Kinyarwanda press accounts, public statements by city officials that can be checked online—to show that my argument is not writerly sleight of hand but is backed up with popular discourse.

Bannyahe is an exemplar of this chapter's central argument: that Kigali residents restore their city from the destruction of sustainable urbanism not through a drive to "resist" but by hacking into and hijacking state functions through collective processes of unauthorized repair—what everyone in Kigali calls akajagari, and what I call, inspired by insights from African cyberpunk, *punk urbanism*.[6]

The punk label has been applied to everything from music and fashion to video games; my concern here is with the literature and visual media of punk—specifically, cyberpunk and its close, if rarely acknowledged, relationship to African urban theory and repair studies.

PUNK URBANISM

Like other varieties of punk, cyberpunk is more than a style. It is a subgenre with a perspective that is anticonformist, celebrating nonalienated modes of production against the alienating market logics of corporate capitalism. Cyberpunk is often characterized as originating in Euro-America (see Csicsery-Ronay 1988), but cyberpunk spaces have always existed in Africa and elsewhere in the Global South, and the subgenre owes many of its narrative conventions to postcolonial cityscapes. For example, the setting of William Gibson's cyberpunk classic Bridge Trilogy is the near-future San Francisco Bay Bridge, but its inspiration was Hong Kong's Kowloon Walled City, which, until it was demolished in 1994, was one of the longest-running self-sufficient squatter communities in human history. To make his bridge community, Gibson takes Kowloon out of its original postcolonial context and stamps the neighborhood onto a US cityscape, presenting Hong Kong's past as California's future.

The worlds that US cyberpunk writers offer as fictitious futures in US cities are already present in Kigali and many other African contexts. The Kinyarwanda/Kirundi cyberpunk musical film *Neptune Frost* (2021) situates a future community of gender-nonconforming hackers in present-day Rwanda. Filming on site in Kigali and neighboring towns, the film's directors, Anisia Uzeyman and Saul Williams, use the present built environment of Kigali as their futurescape. In the film's production, Uzeyman and Williams draw on and highlight the virtuosity of existing repair practices in Kigali while portraying the conditions of inequality that force people to engage in this work. Every set is built from discards in ways that offer Kigali's present as an image of a planetary future: cyborgs with limbs made of e-waste that is dumped in Africa and built environments assembled from materials that have been broken down and repurposed; jerry cans, beer bottles, old televisions made into architectures. Uzeyman and Williams's hackers are pursued by, but eventually evade, global extractive economies that are propped up by a fictional but easily recognizable hybrid corporation/police state.

The backdrop to Lauren Beukes's classic South African biopunk novel *Zoo City* (2010) is the inner-city Johannesburg neighborhood of Hillbrow. Like *Neptune Frost*, nothing needs to be adjusted in Beukes's fictitious city because Johannesburg is already the future. The actual neighborhood of Hillbrow and Beukes's *Zoo City* are maintained by a collection of outcasts who are excluded from everywhere else in Johannesburg. In both today's reality and the fictive world, a once whites-only neighborhood (Hillbrow) is occupied by Black residents—mostly non–South African immigrants and refugees. Unlike AbdouMaliq Simone's (2004b) account of Hillbrow as "people as infrastructure," in which social relations are invented on the fly in the service of mere survival, Beukes's *Zoo City* highlights what people do with the material infrastructures left behind by apartheid. Much of Hillbrow consists of "hijacked" buildings in which the technologies of apartheid have been taken over by the very people that architecture was designed to keep out. Beukes's characters, like the actual residents of Hillbrow, have accessed a space they were excluded from, but in converting the shells of former apartheid-era luxury apartments into low-cost rentals, they are also a stigmatized reserve of cheap labor for the rest of the city.

What makes Africanfuturist punk like *Neptune Frost* and *Zoo City* so useful in understanding spaces like Bannyahe is how they elevate the essential labor of repair and maintenance—hacking infrastructures, hijacking buildings, de-wasting trash by converting it to useful things. This work foregrounds the essential labor of repairing and maintaining in the aftermath—labor that much scholarship in repair studies points out is relegated to the background in conventional social theory because of a "productivist bias" that privileges design and novelty while devaluing the people and the vital, collective labor of maintenance (Jackson 2014, 226; Corwin and Gidwani 2021, 1). In addition, Africanfuturist punk like *Neptune Frost* and *Zoo City* attends to the embodied experience and expertise of people who are so often dismissed, silenced, or treated as mere victims of state violence rather than as political agents with the environmental consciousness and technical capacity to build their own city.

Much Africanfuturist punk that deals with repair and maintenance aligns with a strain in repair studies that understands the process of breaking things down and rebuilding them as challenges to top-down planning and control. Other work on repair and maintenance shows that they are just as easily used to support and legitimate technocratic authority (Doughty 2020; J. Doherty 2022, 54; Barnes 2017). The unauthorized repair in Bannyahe and the street economies it supported shows that these two strains are not mutually exclusive. Bannyahe residents built an unauthorized neighborhood and frustrated ministry-level officials by supporting, rather than resisting, a diffuse range of lower-level urban authorities who were then used against efforts to displace them.

Unsurprisingly, much postcolonial African punk centers spaces in Africa, usually real cities where, as Mathew Omelsky points out, "the implicit weight of (post) colonial history, violence, and exploitation [lies] beneath the surface" (2014, 39). As Jessica Dickson argues in her brilliant analysis of *Zoo City*, African punk (like postcolonial sci-fi in general) "usurps the tools of cognitive estrangement wielded by western imaginaries and redirects them" (2014, 68). It is a perspective that, Nalo Hopkinson famously argues about postcolonial sci-fi in general, takes "the meme of colonizing the native and, from the experience of the colonizee, critique[s] it, pervert[s] it, fuck[s] with it, with irony, with anger, with humor" (Hopkinson and Mehan 2004, 9).

As I show below, from this perspective of critiquing, perverting, and fucking with the utopian vision of the master plan, and by repairing neighborhoods that were supposed to remain broken, Bannyahe was decidedly punk. What makes Bannyahe and the other akajagari Kigali neighborhoods like it punk is what makes the refuge in Uzeyman and Williams's *Neptune Frost* and Hillbrow in Beukes's *Zoo City* punk: it was an act of intentional and collective unauthorized repair by people who were supposed to be erased from the future—one that also usurped the utopian estrangement wielded by the Kigali City Master Plan in the aftermath of sustainability-induced destruction.

A Kinyarwanda concept that saturates popular discourse encapsulates these activities while capturing the contradictions of repair work: *akajagari*. *Akajagari*, like *punk*, is an amorphous term. Sometimes it is defined pejoratively by non-Kinyarwanda speakers as "urban disorderliness especially among the lower classes" (Goodfellow 2022, 201). But there are other, more precise definitions that are used in the specific context of contemporary urban management in Kigali. For example, when I asked Mr. Bimenyimana, a middle-class Bannyahe landowner, how to define it, he responded, "Akajagari? Right [*bon*]." Then, after a pause, he replied, "You can say it means to build in a way that they [city authorities] don't authorize [*uburyo batameye*, they don't agree with]" (recorded interview, September 26,

2014). City authorities agree with Mr. Bimenyimana's definition. They define *akajagari* as any space or practice that is not officially authorized by upper-level city managers. *Kubaka akajagari* (to build akajagari) is to engage in unpermitted construction or repairs. *Gucuruza akajagari* (to sell akajagari) is to engage in Kigali's street economy. The overlapping means and relations of generating both forms of akajagari gave shape to Bannyahe.

UNAUTHORIZED REPAIR

I was born in 1967 in Kiyovu cy'abakene . . . you know? I cannot say it was a particularly beautiful place [ahantu heza] on the poor side of Kiyovu [laughter] . . . but life was good, you get it?
—RECORDED INTERVIEW WITH MR. MUTONI, MAY 2, 2014

When I interviewed him in May 2014, Mr. Mutoni and I were sitting in his small but tidy living room in his two-bedroom bungalow in Bannyahe. His home was well ventilated, connected to electricity, and just a few meters away from a municipal water pump. He sent his two children, who were, in his words, harassing the *muzungu* (white boy [me]), out to play with their neighbors' children. He then shut off the television, which was playing a Kinyarwanda film from a flash drive.

"They wanted us to move to those tiny houses in Batsinda!" Mr. Mutoni said. Then he laughed. "They [the city] wanted to break akajagari! We made more [laughter]" (recorded interview, May 2, 2014). Mr. Mutoni then walked me around his house. His corrugated roof, his windows, and his doors were all salvaged from the demolitions in Kiyovu cy'abakene. The rest of the house was built with amatafari ya rukarakara made from clay in the wetlands below and sealed with karabasasu.

Bannyahe was not the only place where evictees from the demolitions in the urban core went. The neighborhood was connected to a citywide network of akajagari spaces. One hundred and twenty-six families went to Batsinda, got out as soon as possible, and then repaired their homes in the Dushenye area of Kagugu (see chapter 4). Others moved across the wetlands to Kosovo and lived there through the master plan–induced floods described in chapter 1. Still others went to Sodoma ("Sodom") in Gikondo and Dobandi ("for gangsters") on the edge of Nyarugenge.

Before the evictions described in chapters 1, 2, and 3 began, the area that would become Bannyahe was located on the city limits; it had few inhabitants and fewer amenities. It was, in the words of long-term residents, "bad country" (*igicaro kibi*), seen as rural because it was not connected to the city's hydraulic, electric, or transportation grids.

But Bannyahe would not be bad country for long. As the city's bulldozers pushed people out of the urban core, evictees like Mr. Mutoni broke down their

FIGURE 14. Satellite images of Bannyahe in 2000 (*top*) and 2013 (*bottom*), when this research began. Source: Google Earth, Digital Globe. These images support Bannyahe residents' recollections that the neighborhood was little more than a few scattered farms during the postgenocide population boom. Bannyahe was built largely as a response to demolitions resulting from the OZ plan and, later, the Surbana plan by people who ignored policies against unauthorized repair. In the bottom right are three stabilization ponds used for sewage. Above these ponds and to the right is the wetland agriculture.

homes into component parts. They took the materials they could carry and rebuilt their homes. New arrivals to the city also found housing in Bannyahe. Fed and clothed by discount street economies, the new residents rebuilt the homes, markets, churches, and bars that once served Kiyovu cy'abakene, Kimicanga, and Gacuriro (see figure 14).

And, like Sodoma ("Sodom"), Dobandi ("for gangsters"), Kosovo, and Dushenye ("they demolished us"), Bannyahe earned its own famous moniker, a sly scatological comment on Kigali's brand image as the "cleanest city in Africa." Technically, Bannyahe/Bitma is a question. It means "Where do they piss/shit?" This naming is what Situationists and punks call *détournement*: hijacking dominant ideology by ironically engaging with Kigali's branding as "the cleanest greenest city in Africa."

The largest issue with Bannyahe, for both scholars of the Rwandan state and the City of Kigali, was that it was not supposed to happen. By the time Mr. Mutoni and most of his future neighbors moved to Bannyahe, the Ministry of Infrastructure and the City of Kigali had written permitting laws and building codes and released sector-level zoning maps, all directed at preventing Kigali residents from repairing the city from master plan demolitions. These regulations were supposed to be enforced on the ground through what Hugh Lamarque, in his work on community policing in Rwanda, calls "a remarkably efficient information gathering apparatus" (2020, 159).

That apparatus, which continues today, begins at the community level (umudugudu). Abanyerondo, night patrols made up of mostly men and some women from the area, are supposed to keep watch for unpermitted construction or repair. They report to community leaders and police liaisons who, in theory, inform sector-level authorities about unauthorized buildings and repairs. Serving as a check to abanyerondo are sector-level local defense units (called *abadasso* since 2013), who also patrol neighborhoods—day and night—and report unpermitted building to sector-level offices that then order the destruction of unauthorized buildings. Sector authorities report to district mayors, who report to city authorities. Since 2008, abanyerondo and abadasso have also been tasked with arresting street traders and sending them to extralegal detainment facilities (see chapter 2).

Policy-focused scholars have debated whether this distributed network of security agents, which, in theory, keeps Rwandans under close state surveillance, is an example of authoritarian overreach (see Longman 2017; Purdeková 2015) or efficient community policing (Lamarque 2020). But these conflicting analyses reach a striking level of agreement on one point: most scholars of the Rwandan state assume that the security apparatus *works*. It works in the sense that the state apparently does prevent Kigali residents from constructing houses and making unpermitted repairs by writing policy or painting Xs on buildings (see Hudani 2024, 108; Sommers 2012, 218). Throwing street traders in extralegal detention facilities has apparently "led to a shrinkage of informal activity from the city centre" (Goodfellow 2022, 225). But Bannyahe, like Dushenye and many other areas of the city that were built after this network of surveillance was installed, and the visible growth of the street economy offer a different picture: the City of Kigali does not have control over its own surveillance networks, let alone the "informal activity" that it seeks to police.

"Of course, they will demolish us," Mr. Mutoni told me from his home in Bannyahe three years before the neighborhood's demolition was announced, "and then we will take our poles and we will take our roofs and we will go build more akajagari somewhere else."

Ministry-level government representatives also characterize the building of Bannyahe as unauthorized repair from master plan destruction. In a 2018 address to Bannyahe residents, recorded by the Kinyarwanda press, the minister of local government expressed his frustration with how residents had simply rebuilt the neighborhoods the city tried to demolish. "We [the government] don't move people [like you] from akajagari just so you can build more akajagari!" (Kayiranga 2018, 3). After encouraging Bannyahe residents to give up their neighborhood and move into foreign-owned housing estates, the minister concluded his talk with "I trust that you now accept/understand."

The last line was low-hanging fruit.

"Nearly all of the meeting attendees," the Kinyarwanda press reported, shouted back "in unison 'We don't accept!'" (Kayiranga 2018, 4).

The largest issue, from the perspective of the minister that day, was that city residents had rebuilt akajagari faster than municipal-level authorities could tear it down. They did this by hacking into and hijacking the very systems aimed at preventing Kigali residents from repairing their neighborhoods after master plan destruction.

HACKING "STATE INFRASTRUCTURAL POWER"

Neptune Frost (2021) is about, among many other things, the Matalusa Kingdom,[7] a future community of hackers who have built a refuge from a fictional corporate government by breaking down discards and putting them back together as new technology (prosthetic limbs made from e-waste), futuristic fashion (outfits assembled from discarded fabric), and shelter (houses made from jerry cans, bottles, and TV screens). In the film, the Kinyarwanda verb *kwinjera*, which generally means "to enter," is translated in another way it is also used in Kinyarwanda: "to hack." This translation works by returning precision to a term that has long been watered down by neoliberal appropriations.[8] In punk and in akajagari, to hack is to make an unauthorized entrance without destroying the infrastructure that holds the system in place. Hackers manipulate the weak links in a system's network to bypass firewalls and redistribute access to services and technologies that infrastructure is designed to block. Once inside, they rewrite the program, creating alternative pathways, repurposing the existing infrastructure so that services can be accessed by others.

In Kigali, the infrastructure that hackers seek illicit access to is a central tool of state regulation. As Tom Goodfellow writes, authority in Kigali is "accompanied by a strong emphasis on securitizing urban space through new forms of top-down social organization and surveillance, facilitated by the capillaries of infrastructural power embedded in Rwandan society that the RPF were able to capitalize on" (2022, 147). Goodfellow calls these surveillance systems "state infrastructural power." In theory, the RPF-run state operates a network of total control at every level of city life. The "infrastructure" in the state's infrastructural power includes both the city's hardware of pipes, streets, and wires, which is ostensibly under RPF control, and the less material networks of surveillance, like security forces that operate in every aspect of society.

Bannyahe residents connected their neighborhood to the city by capitalizing on this infrastructure. In the process, they undermined the city's ability to prevent akajagari by pitting different government projects against each other.

"*Reka da* [not at all]!" Mr. Murenzi responded with a laugh when I asked him if they had electricity or water in 2002, the year he moved to Bannyahe from a rural area outside the city. "It was like living in a forest then!" (recorded interview, September 26, 2014). Mr. Murenzi's neighbor, Mr. Mutiyimana, who moved there when his neighborhood was destroyed to make way for the convention center, agreed. "When the [master plan] demolitions began," he said, "we all came here from all over [*mpande zitandukanye*, literally 'different directions'] . . . Kiyovu, Kimicanga, Gacuriro, every day there were new houses! But we were still living in the dark!" (recorded interview, April 17, 2014).

While Mr. Murenzi and Mr. Mutiyimana agreed with city authorities that their neighborhood was akajagari, they were also both, in theory, complicit in the structures of authority designed to suppress akajagari.

In 2014 Mr. Murenzi was one of the people tasked with preventing akajagari, as one of Bannyahe's abanyerondo. But rather than report his neighbors (and himself) for illicit building practices, Bannyahe's abanyerondo looked the other way: "then . . . [in the 2010s]," Mr. Murenzi recalled, "you would come home to find five new houses up! [laughter]" (recorded interview, September 14, 2014).

Evictees from master plan demolitions in the urban core arrived in Bannyahe with roofs and poles salvaged from the bulldozers. They purchased plots from landowners using the system of inzoga z'abagabo described in chapter 2 and then registered their land during the mass registration program described in chapter 3. But they had no connection to the city, which meant no way of making money and no services. As in previous generations, they built the neighborhood not by accepting "bare life" but by building the state.

In 2009, as Bannyahe's population grew, Mr. Mutiyimana, Mr. Murenzi, and three other Bannyahe landholders started a community savings association in an effort to connect Bannyahe to water and electricity. Their plan was to pay the utility company to connect them to the city's water and electricity grid. "We studied the utility company, you know?" Mr. Mutiyimana laughed. "To learn what they wanted."

A year later, in 2010, the national public utility company was converted into a public-private partnership called EWSA (Electricity, Water, Sanitation Authority). That year, the Bannyahe savings association that Mr. Mutiyimana and Mr. Murenzi started contacted EWSA management requesting electricity hookups. The neighborhood was, after all, surrounded by the best-serviced areas in the city. Technically, all it needed to get connected was to extend the infrastructural base from its wealthy neighbors. The utility company responded by sending a technician out to survey Bannyahe. The answer came back: no.

"They [EWSA] were helping the mayor fight akajagari," Mr. Mutiyimana explained, "and we—we are akajagari, you see?" But the utility company couched its refusal as a technical rather than a political problem. The savings association tried again with a different approach, appealing to EWSA's recent transformation into a public-private partnership and offered to resolve the contradiction between the utility company—whose leadership was then under considerable government pressure to connect as many city residents to electricity as possible—and city leadership, who were trying to prevent the extension of services to akajagari neighborhoods. Mr. Mutiyimana described the back-and-forth: "At first they [EWSA] told us, "No, it's not possible [to install infrastructure]," and if you look at the neighborhood, it was true. There was no room for a truck to bring in electricity poles and nowhere to install the poles . . . but slowly we kept going back to them to ask questions. How many meters of space do we need? Like that." A year after its initial request for electricity, the association went back to EWSA. "We told them," Mr. Mutiyimana said, "we would pay them for the poles and install them ourselves. That was 2010 when we did it" (recorded interview, August 22, 2014).

In 2010 Bannyahe residents paid the utility company RWF 2 million for the hardware they needed to connect to their neighbors' electricity grid. EWSA, following a state program that subsidizes community infrastructural upgrades, covered the other RWF 2 million. The utility company dropped the electricity poles on the side of the road. Using umuganda, the monthly sanctioned day of community service, Bannyahe residents tore down several houses to allow installation of the poles in a grid formation that met EWSA's requirements for one pole every five hundred meters. They then rebuilt the houses, using amatafari ya rukarakara.

Mr. Mutiyimana invited me outside his house to illustrate this process. We walked to the path just outside of Bannyahe, which offered a good view of the neighborhood as it sloped into the wetlands. "For example," he said, pointing to an electricity pole sticking out above a roof that had two colors, "you see that house there? There is that panel in the roof that you can see does not look like the rest—we had to tear down half of that house and then rebuild it, to put the pole in. Then we put the house back together and repaired the roof in one day!" (see figure 15).

FIGURE 15. The lower section of Bannyahe, from the path leading to it. Note the straight-lined precision of electricity poles installed every five hundred meters over the zigzag base of the neighborhood. Also note how nearly every structure is built with amatafari ya rukarakara sealed with karabasasu. Photo by Samuel Shearer, 2015.

The following year, in 2011, the association repeated the process to get connected to municipal water, paying the utility company another RWF 2 million for fixed capital—the pipes and the materials for municipal pumps. More structures were torn down and rebuilt to accommodate the new infrastructure. Residents built the water kiosks, helped dig the ditches for pipes and drains, and put the neighborhood back together after installing the infrastructural base.

Installing water and electricity was about more than turning the lights on, as Mr. Mutiyimana explained. It was about reconnecting and reentering the city by creating a fully serviced neighborhood. Without electricity and water, their houses were just empty shelters. With services intact, landowners who were evicted from the urban core could rebuild the stock of African-owned affordable rentals, cafés, bars, and shops lost in the demolitions and reconnect to the rest of Kigali. "Before [water and electricity]," Mr. Mutiyimana recalled, "You could not find money here . . . I have these two plots, now I rent rooms, because people want to live here. Before you couldn't do anything! If you don't have electricity, you cannot sell anything, no Fanta, beer, nothing! No one will come to a bar at night if there are no lights. But now, there are jobs, shops, you can work."

Scholars of infrastructure and services in other African cities observe how expert technicians in Dar es Salaam, Nairobi, and Johannesburg tinker with pipes, improvise with wires, and break meters to gain illicit access to services that they would otherwise be denied (see Degani 2022; Schramm and Ibrahim 2021; Von Schnitzler 2016). But Bannyahe's landowners engaged in hacking of a different sort: they crossed a barrier created by one state project—the master plan—by paying for and putting together a licit connection that would benefit another: the utility company. In the process, they certainly supported and legitimated ruling party authority. By installing pipes, poles, and meters in the neighborhood, they connected Bannyahe to both the city and the state that governs it while using government programs like umuganda. They then hijacked these processes to use them against efforts to align the city with the tabula rasa envisioned in the master plan.

Once connected to the city via water and electricity Bannyahe redirected its economic activities. The neighborhood went from a primarily farming community to a new urban center of economic activity that spread to the rest of the city, overwhelming city authorities with illicit street economies that take over the urban core every day.

UNAUTHORIZED MAINTENANCE

Every morning before sunrise, Ms. Kayatesi, a former resident of Kiyovu who moved to Bannyahe after the 2008 demolitions, and dozens of other abazunguzayi (mobile street traders) from Bannyahe walked the seven kilometers to Nyabugogo transport hub. There, at Nyabugogo's wholesale produce market, they would join hundreds of other abazunguzayi from Dushenye, Dobandi, Sodoma, and elsewhere. Ms. Kayatesi and the women she worked with would pool their cash to purchase "third-quality" produce that is near the end of its shelf life and therefore marked down. Pooling their resources to buy bulk gets them a further discount that allows them to reach their client base: people who can afford to buy only what they will eat that day. Ms. Kayatesi and the other three women then divide their stock, disaggregating their bulk purchases into rations that will sell for RWF 100–200 (about $0.25), and hit the streets with woven baskets full of produce to meet the crowds of pedestrian commuters.

Working together with large numbers of other abazunguzayi, Ms. Kayatesi and her colleagues take over high-traffic areas of the city center. They push into Marato (Marathon, named after the daily street raids), which sits at Nyabugogo transport hub. They take over the central business districts of Commercial and Mateus, and they occupy the streets outside of the high-end downtown shopping centers. The street economy overwhelms the urban core as abazunguzayi overtake spaces that have been rezoned in the Kigali City Master Plan for high-end retail, housing, and conference tourism. Ms. Kayatesi and her colleagues push through the crowds, selling discount produce from baskets carried on their heads or laid out at their feet, while men and women who sell secondhand clothing display their wares on their arms.

I asked Ms. Kayatesi in an interview in 2014 how a diverse group of people engaged in varying forms of street commerce came to be categorized as a single population—abazunguzayi. "We have no address [*nta address dufite*]," she replied. This reply speaks to what many street traders see as the defining feature of their profession: mobility. The root of the word *abazunguzayi* (singular *umuzunguzayi*) is the verb *kuzunguza*, to wander, twirl, or shake. Asking an umuzunguzayi for specifics on how they do their job will elicit an assortment of verbs that index nonlinear movement. Straight money? That doesn't exist. If you want to make and keep cash on Kigali's streets, you need to zigzag, cut, run, erupt, and fly from one location to the next. You must always be prepared to exit stage left, discount a product to make a quick transaction, or part with your merchandise at a moment's notice.

Zigzagging is a central concept in the maintenance of street space in Kigali. *Gukatakata* (to zigzag) is used to describe physical movement that maintains popular control over the streets—crowding in and hanging out, but also dodging or fighting in security raids. But *gukatakata* also designates the tacit knowledge that one needs to make it in the street economy: the ability to manipulate multiple temporal and spatial registers and navigate the city's illicit economies.

This sense of zigzagging as a market strategy is certainly not unique to Kigali. It seems to be a general condition of urban life in many cities across Africa. In his work in Harare, Zimbabwe, for example, Jeremy Jones observes the common usage of a similar chiShona concept, *kukiya-kiya* (also "to zigzag"), "that suggests cleverness, dodging, and the exploitation of whatever resources are at hand, all with an eye to self-sustenance" (2010, 286). What makes this concept significant to Jones, however, is not that it indicates that some people hustle, while others don't, but that in contemporary Harare, "the entire economy has come to be defined as *kukiya-kiya*" (Jones 2010, 286). Writing about Oxford Street in Accra, Ato Quayson (2014) also observes zigzagging as a general condition of urban life. On Oxford Street, hawkers and their stalls have taken over the sidewalks, pushing pedestrians out into the street, where they compete with motorcycles, tro-tros, and automobiles. The physical movement of pedestrians, hawkers, and motorists zigzagging around and into each other produces a theater of social interactions in which actors collide and trade insults and jokes in front of a participating audience of onlookers.

Abazunguzayi understand zigzagging as Jones and his interlocutors use the term—as the tacit knowledge necessary to hustle what Jane Guyer (2004) calls a "marginal gain" from the streets. And, as Quayson does, they also use the concept to describe corporeal movements and interactions that carve urban spaces out of limited built environments. But unlike in Harare, Accra, and Kinshasa, where state authorities—when they are present at all—zigzag along with the streets, abazunguzayi and their clients employ zigzagging to maintain popular access to Kigali, its central business district, and the transportation hubs after authoritarian attempts to evict them from the city center. They do this not through small "minor acts" but through a consciously organized collective politics of anticipation. Working in large numbers, through labor that is essential to the maintenance of the entire city, they know that city authorities will always, and hopelessly, fail to evict them from the streets because the rest of the city depends on them.

The numbers game

Maintaining ownership over the streets against municipal aspirations to clear them in the service of the plan, Ms. Umulisa—an umuzunguzayi who moved to Kagugu after her home in Kimicanga was demolished—explained is a numbers game (her word, *umukino*, "game"). The game involves a critical mass of abazunguzayi working together to meet the needs of a consumer base that includes the most essential labor in the city—so it can be exploited—by using carefully designed financial institutions that offer durable safeguards against risk and enable vendors to control the street economy and the spaces where it thrives. The thousands of women abazunguzayi who control the produce trade, for example, agree that between RWF 5,000 and 10,000 (about $7–$13) is an ideal amount to begin the morning with. With low starting finances, however, manipulating bulk and quality is essential. A few colleagues can pool their money and talk a wholesaler into letting them take several kilograms of "third and fourth" quality tomatoes—tomatoes with only a day of shelf life left. They then split up the haul into, say, two-kilogram rations.

Now each seller needs to unload her two kilos. She makes her gains by manipulating quantity and breaking the stock down further, disaggregating the smallest weighted unit of wholesale (the kilogram) into affordable daily rations. Depending on the size of the individual fruit, these will be small piles of two, three, or four tomatoes that sell for RWF 100 ($0.10) each. The profits should be between RWF 1,000 and 2,000—10 percent. After she pays her daily *ibimina*,[9] she takes the rest to the streets and picks up her own hundred-franc piles of food for the day.

There are, of course, other ways to work. More established abazunguzayi, for example, have relationships with market vendors, which give them access to better-quality goods, higher returns, and stock on credit. The men and women who run the secondhand clothing trade discount out-of-fashion or lower-quality secondhand clothing on the streets, moving surplus out of shops and market stalls; many work on consignment with market vendors to move the clothes that do not sell in the market.

This well-designed system for manipulating time, quality, and quantity by breaking down merchandise from larger to smaller units; by stepping in to prevent the wasting of produce that is about to go bad; and by providing an alternate outlet for merchandise that can't be sold at the markets is underwritten by financial associations that keep capital in circulation. Most Kigali residents participate in ibimina, collective savings associations. But ibimina are necessary for abazunguzayi. Daily returns in the discount produce trade are predictable but low, about RWF 800–2000 a day ($1–$2.75 in 2015). In addition to the expenses of daily sustenance, large payments such as rent, school fees, and monthly credit lines will come due. Ibimina helps participants to manage these expenses. For example, an ibimina group may meet once a day for a month; its thirty or so members will decide together what each member should contribute and how the funds will be distributed. Every member may contribute perhaps RWF 500 a day, creating a total pot of RWF 15,000 each day. On the first day, the names of all thirty members are put into a hat, and one person wins the entire pot by lottery. Everyone returns the next day and repeats the lottery, minus the name of the person who won on the first day. This process is repeated each day. Only one person in the group will have to wait the full time it would otherwise take to accumulate RWF 15,000, with everyone else accessing interest-free credit to pay their bills. The trick is to belong to as many ibimina as possible so that big payouts are constantly coming in before the bills arrive.

In the absence of formal regulation, ibimina work only in the presence of a sense of community, shared responsibility, and coherent ethics.[10] These community savings arrangements are not in any way unique to Kigali. They are part of what the Kenyan geographer Mary Njeri Kinyanjui (2013, 2014) calls "solidarity entrepreneurialism"—business ethics based on communal care and collective well-being rather than competition and individual profits. Drawing on her own work with women street vendors in Nairobi, Kinyanjui argues that solidarity entrepreneurialism constitutes an alternative mode of enterprise, one that is outside the large corporate logics of capital accumulation (Kinyanjui 2013, 150). To be sure, Kigali's street economy operates outside of, and in opposition to, the corporate logics of the master plan and the ambitions of zero-tolerance policing that support it. But solidarity entrepreneurialism, like unauthorized repair, hijacking, and the hacking of infrastructure, is effective not because it is alternative in some way, but because the street economy as a discount economy is essential to maintaining all of Kigali.

Street politics

The systems of disaggregation, discounting prices, and combining resources in the streets are, to be sure, about collective survival. It is the politics of survival that give abazunguzayi the upper hand in their struggle with the mayor's office over the streets. Most people who live in Kigali either directly or indirectly depend on the street economy to survive in the city. The largest, most essential, lowest earning, and most dangerous work in the street economy is the produce trade, which is controlled by women who follow the same logic that Mr. Mutoni and his neighbors used to rebuild their homes in Bannyahe: they break down what is already there into components that can then be distributed.

To secure their foothold in the streets, women abazunguzayi capitalize on the processes of decay. They take goods, often third-quality goods that would otherwise go to waste, and move them quickly, serving the city's majority consumer base that lives outside the pricing limits of regulated trade. In doing this work, Ms. Umulisa, Ms. Uwase, Ms. Kayatesi, Ms. Ashimwe, and the thousands of other abazunguzayi feed the city's essential workers. For example, *abafundi*, skilled construction workers, make RWF 5,000 a day (a little over $5) but rarely work every day of the week and sometimes need to stretch a week's wages through several months of unemployment. *Abayedi*, unskilled day laborers, make only RWF 2,000 a day and, like their skilled coworkers, often go weeks without work. Sanitation cooperatives pay street sweepers (also mostly women) RWF 20,000–30,000 per month (less than $1.50 a day) to keep Kigali's famously clean streets spotless. Waste collectors and sorters make RWF 50,000 per month (less than $2 a day). Public school teachers start at RWF 60,000 per month ($2.50 a day) and sometimes go months without a paycheck.[11]

On RWF 2,000 a day, purchasing produce by the kilo in Kigali's regulated markets, where taxes, rent, quality, and quantity are built into the price, is simply out of reach. But disaggregated and discounted goods that are at the end of their shelf life, sold on the streets, are affordable. Those goods allow buyers to make it through another day, to go to work maintaining the city, cleaning its streets, and teaching its children. Thus, although the women who run discount street economies are cited by city authorities as antithetical to security and associated with vaguely "unhygienic" practices, they are crucial to the maintenance of the entire city. They reduce waste loads and engage in collective organizations that feed the city. At the same time, this discounting work also maintains the unequal relations that sustain capital by keeping wages low, cheapening the labor and lives of people that city managers and high-end construction firms depend on.

These contradictions are reflected in the careful balance that security forces maintain between carrying out orders from higher-level authorities and allowing unauthorized street economies to persist. The mayor's office and Rwanda National Police have responded to the street trade by sending abanyerondo out to raid the streets, and many abazunguzayi spend long periods detained in extralegal detention facilities when they are caught in raids. But just as often, the security units tasked with arresting street traders walk right past crowds of abazunguzayi and their clients, pretending not to see them. Abanyerondo like Mr. Murenzi (above) come from the same neighborhoods as abazunguzayi like Ms. Kayatesi and themselves depend on the street economy as a source of both rents and discounted goods. In addition, security agents are highly incentivized to ignore the street trade by the threat of violence.

When they are arrested, abazunguzayi are detained for days, sometimes weeks, at Kwa Kabuga, a warehouse that has been converted into an extralegal detention facility. The conditions of overcrowding, violence, and disease at Kwa Kabuga are so bad that it has been the subject of several Human Rights Watch reports (see, e.g., Human Rights Watch 2006, 2015). But as bad as Kwa Kabuga is, few abazunguzayi speak about it in somber tones. Like the name Bannyahe, Kwa Kabuga is also a joke—a violent manifestation of the municipality's hopeless failure to enforce its own laws or exercise any real authority over the streets. As Ms. Umulisa and Ms. Uwase told me one day in 2015, "No, listen! The *abapandagari* [cops][12] might take one of us, right? But we are always more! One [umuzunguzayi] goes down and a hundred take her place, you get it [*hapfa umwe ijana rikanjira*, literally 'one dies and a hundred enter']? [laughter]" (recorded interview, April 13, 2015). Ms. Ashimwe, an umuzunguzayi who lived in Bannyahe, echoed the sisters in a separate interview: "They [the police] catch you, and you know someone is coming right behind you [to take your place] [laughter]. They lock you up today, you will return tomorrow" (recorded interview, October 10, 2014).

Ms. Umulisa, Ms. Uwase, and Ms. Ashimwe's blasé—and very punk—attitude toward the violence of extralegal detainment and street raids matches the exasperation city authorities express toward the street economy. For example, in 2017, just a few months before he announced the plan to demolish Bannyahe, the then mayor of Kigali declared to Kinyarwanda media, "Abazunguzayi and akajagari have continued to overwhelm the City of Kigali" (Niyomwungeri 2017).

Many abazunguzayi agree. Attempts to prevent akajagari from taking over the central business district through policing have been, as Ms. Umulisa says, remarkably ineffective. A raid on the city center might result in the arrest of maybe five to ten abazunguzayi, while the rest get away. Thirty minutes later, everyone returns to the streets. And abazunguzayi are not alone in defending the street economy. Others step in to defend it, too.

For example, on May 7, 2016, twenty-seven-year-old umuzunguzayi Théodosie Umwamahoro was beaten to death by two *abadasso* (district-level security guards) in front of a large crowd of street traders, customers, and onlookers during a raid on Nyabugogo transport hub. The murder of a single mother for the crime of selling produce captured the attention of media, leading to an impromptu protest that shut down the transport hub (see Ukwezi TV 2016, https://www.youtube.com/watch?v=JRlovkD2-pk, for footage of this protest).

The timing of Umwamahoro's murder and the subsequent protest could not have been worse for the City of Kigali: just a week before international delegates were set to arrive in Kigali for that year's World Economic Forum on Africa.[13] The total breakdown in local security in the face of the protest forced the city to call in the Rwanda Defence Force and the minister of gender to get things under control. The minister set up microphones, allowing women abazunguzayi to publicly voice their grievances about state violence against the street economy and corruption, topics usually censored in the Rwandan media.

Policing is not the only strategy city authorities have used in efforts to eliminate the street economy. They have built markets and offered rent-free retail space; they have organized abazunguzayi into cooperatives; and they have tried training women abazunguzayi in new crafts. State institutions, particularly the Ministry of

Gender, are uncomfortable with the amount of violence that women abazunguzayi are exposed to through street raids (see Shearer 2020a). But none of these programs have reduced the street economy or spaces like Bannyahe that it supports.

More recently, large numbers of Kigali residents have come to the defense of the street economy by simply beating security units down, providing more incentive to leave the street economy alone. That is what happened on August 29, 2022, in broad daylight just outside the former Nyarugenge Market in the colonial core.[14] On that day, when abanyerondo tried to raid the city center, a small group of abazunguzayi and a large group of other residents jumped into the back of a community police patrol truck and attacked the abanyerondo until they released the abazunguzayi they had just arrested. After they let the street traders go, the community police unit tried to get away from the crowd that had surrounded their vehicle and were throwing bottles and other trash at them. The event was caught on a smartphone camera, and the footage was sent to the Kinyarwanda news agency BTN, which uploaded it to YouTube (BTN TV Rwanda 2022; see https://www.youtube.com/watch?v=-mxaeQphtkw&list=PPSV).

The community police unit did escape on August 29, but another patrol, seven months later, had a more tragic outcome. On March 7, 2023, a raid on the street economy at the downtown CHIC complex became deadly when abanyerondo tried to chase down a woman who had just escaped from custody. In pursuit, the abanyerondo ran into a crowd of other street traders, who swarmed them. In the ensuing fight, one of the patrolmen was knifed to death (Ngabonziza 2023). Even as they acknowledged the death of the patrolman as unfortunate, many Rwandans also came to the defense of the abazunguzayi who killed him. Anonymous commenters responding to online reports of the incident praised the abazunguzayi as "heroes" for fighting back against the abanyarondo (see, e.g., comments to BTN TV Rwanda 2023, https://www.youtube.com/watch?v=ekrWzfLqtgA).

These tragic, well-documented violent events demonstrate that the street economy in Kigali is much more than a "silent" and "stigmatized" "economy on the run" (Goodfellow 2022, 227). City authorities—by their own admission, backed up by secondary sources and media evidence—are *unable* to enforce laws against the street trade. This is not because of the creative resilience of an urban subaltern class. It is because abazunguzayi are so crucial to reproducing the cheap labor (and all the attendant inequalities) that keep the entire city together. It is what scholars of repair studies call the "double sidedness" of maintenance work (Corwin and Gidwani 2021, 3) and what shows up in Africanfuturist punk as the internal contradictions of unauthorized repair. The street economy, as an unauthorized discount economy, emerges from and perpetuates the inequalities and injustices that capitalism depends on (see Barnes 2017; J. Doherty 2022). But it is precisely because the streets are essential to the maintenance of the entire city that Kigali authorities fail, hopelessly, in their efforts to evict abazunguzayi from Kigali's future.

Neighborhoods like Bannyahe are also part of the process of maintaining access to the city by being essential to its everyday management. The builders of Bannyahe and other akajagari neighborhoods hack into state infrastructures to provide the hardware for the street economy: access to affordable housing and services and a pathway into the city for abazunguzayi like Ms. Kayatesi. Abazunguzayi, in turn, keep the people who build (and rebuild) the city fed and clothed via discount economies. The ongoing presence of both forms of akajagari—the street economy and neighborhoods like Bannyahe—rebuilding the city from master plan destruction under the nose of a state that has designed and installed an intricate surveillance system is, to be sure, an example of what Steven Jackson describes as repair: "robust lives ... sustained against the weight of centrifugal odds ... get not only broken, but *restored*" (Jackson 2014, 222; emphasis original). But these processes also complicate conventional understandings of how state power is assumed to rule in Kigali in the service of the master plan.

When hijacking, hacking, and the numbers game come together to maintain akajagari against market-driven sustainable urbanism, they offer an alternative future to the visitor-class utopia envisioned by the master plan, what some Kigali residents call going "cowboy."

ANTICIPATING URBAN FUTURES

I first met Mr. Irakoze, father, homeowner, umuzunguzayi, on the streets in downtown Kigali in 2015. He wanted to sell me a giant laminated map of Rwanda, the kind you might find in an elementary classroom. I declined, but he was undeterred. "Ray-Bans?" he suggested, displaying a pair of plastic-framed sunglasses. Two weeks later, thanks to mutual friends in the upper edge of Kimisagara sector on Mount Kigali, I was sitting across from Mr. Irakoze in his recently built wattle-and-daub house in an area where, according to the master plan, no one was allowed to build. He held his one-year-old daughter in his arms as we talked about life in Kigali, working in the street economy, and police raids. He was the fourth person in his quickly growing neighborhood to agree to an interview about what city authorities called *kubaka akajagari* (unauthorized building) and he and his neighbors called *kubaka kugikoboyi* (building cowboy).

Mr. Irakoze had a lot of experience with the Kigali City Master Plan. In 2004 he moved to Kigali from the Eastern Province as a teenager. He found a room to rent near the tarmac road in Kiyovu cy'abakene and found work in the streets as an umuzunguzayi. Four years later, when Kiyovu cy'abakene was demolished, he moved to Lafreshaîre, near Nyabugogo. Then, in February 2014, after Lafreshaîre was demolished, he moved to N'Djamena (named after the capital of Chad) below the Hotel Impala and rented a room there. When N'Djamena was demolished a month later, Mr. Irakoze gave up on renting: "Everywhere I lived would get demolished and they kept moving me [laughter]. I decided to build my own house" (recorded interview, June 6, 2015).

I had to ask: why buy a plot of land where it was illegal to build a house, and where there were no services? Mr. Irakoze reasoned that owning a house, even in an area where it was prohibited to build, anticipated displacement while converting a basic need for shelter into a speculative enterprise. "So [*noneho*]," he went on, explaining the calculus that went into his decision to buy his own plot of land to build akajagari, "if you have your own house, you can demand compensation when it is demolished and then go and build another house somewhere else. When you rent, you are just getting moved around and losing—you are always being pushed down, you know?"

In his 2012 book *Stuck*, the anthropologist Marc Sommers argues that the Rwandan state's strict building regulations, enforced by local security agents, were preventing young men like Mr. Irakoze from building their own homes and moving into adulthood. In a later work, Shakirah Hudani (2024) writes about the ubiquitous *X*s painted on clay houses throughout the city that are supposed to demarcate a building for demolition. "Accordingly," Hudani argues, "the residents are not permitted to repair or rehabilitate their homes as they wait for formal notice to vacate" (2024, 108). In these formulations, abstract policy is given a remarkable amount of agency to police actions on its own. Accordingly, Mr. Irakoze was not supposed to be able to build a house that was out of code after the master plan regulations took effect.

Indeed, the ubiquitous *X*s painted on houses, while certainly intended to work as a sort of low-budget redlining, are the objects of much ridicule—as they are easily plastered over and usually ignored. Other landowners poke fun at the policy through creative, and very punk, acts that critique the plan "with irony, with anger, with humor" (see Hopkinson and Mehan 2004, 9). My all-time favorite is a shop where the owners used a discarded zoning map from a sector office to partially cover the mark that indicates the shop was itself in the wrong zone, simultaneously covering part of the *X* and demonstrating that efforts to "erase" akajagari are little more than vague aspirations of control rather than effective means of governing construction (see figure 16).

In 2015 I asked Mr. Irakoze how he built his home outside of the building codes. "*Kubaka*," he joked, "*ni kugikoboyi* [to build, (you need to) go cowboy—literally, 'to construct is to cowboy']. . . . Start at midnight," he went on. "Weekends are good."

Going cowboy is a distinctly punk maneuver that hacks surveillance networks and then hijacks them, shutting them down momentarily. Not everyone agrees on the source of the figure of the *umukoboyi* (cowboy; plural *abakoboyi*). The most common explanation is that the term originated in the early independence period when both blue jeans and spaghetti westerns were fashionable.[15] The subject *umukoboyi*, an urban cowboy who hustles and lives outside of regulations, is distinct from the figure of the pastoral *umushumba*, a cattle herder or shepherd.[16]

FIGURE 16. One way to cover up an X: with a discarded master plan printout. Name of location blocked out to protect the proprietor. Photo by Samuel Shearer, 2022.

Building his house, Mr. Irakoze explained, was a lot like the street economy. Success requires working fast, not getting caught, and involving as many people as possible. It requires connections with neighbors and deep knowledge of how to construct a building quickly in a way that scrambles the city's surveillance by hacking it through bribes and working in large numbers. It also, like ibimina, requires a level of anticipatory faith that everyone else in the city will also continue to build akajagari.

Mr. Irakoze's approach was fairly typical of the approach that was used to build Bannyahe, Dushenye, and many other neighborhoods from the ruins of sustainable urbanism. First, to get around master plan zoning restrictions, Mr. Irakoze purchased his plot of land from a *kavukire*, a landholder who had disaggregated his large, inherited plot into several smaller ones that he could sell separately to recent master plan evictees. Thanks to the master plan, which has frozen the city's cadastre into the rigid zoning database, it was nearly impossible—both politically and financially—for Mr. Irakoze to obtain official title to his land, let alone a building permit.

So, everyone involved ignored the law and reverted to the previous system of titling, inzoga z'abagabo (described in chapter 2). Mr. Irakoze brought three neighboring families together to witness the transfer of property. Everyone measured the plot. They all agreed on its dimensions, and then they drew its location and approximate shape on a piece of notebook paper. The title was then signed by everyone present. While the resulting document did not convey any official right to build, it did provide a proof of sale (see chapter 2, figure 4, for an image of a title).

Reverting to the old system of titling rather than submitting a permit to the city through the One Stop Center was not ideal for Mr. Irakoze, as inzoga z'abagabo is not acknowledged by city authorities, and he was clear that he wanted an official title. But it does cause surveillance to malfunction. The City of Kigali never gets a record of the sale. Much to the chagrin of urban planners at the One Stop Center (see chapter 3), no one in the planning office knows where the base map of the plan is shifting until hundreds of homes have gone up and it is too late to intervene without the costly spectacle of demolishing an entire neighborhood.

Not getting caught is a key component of building cowboy. Succeeding at evasion means hijacking the state's infrastructural power to avoid Kigali's famous systems of capillary surveillance. The basic strategy at play here is *akantu/akafanta* (a little thing/little Fanta), or small bribes. I asked Mr. Irakoze how he managed to escape surveillance long enough to build his house. "You mean the *abapandagari?*" he clarified. When I asked, he laughed. "It's not *ruswa* [corruption]," he answered. "Just little things, you know [laughter]? You pay *akafanta* [a little Fanta] here, some over there—like that."

The distinction Mr. Irakoze makes between ruswa (corruption) and akantu/akafanta (little thing/Fanta) is important. While ruswa indicates a moral digression and is used to describe everything from graft to sexual harassment, paying a little something/Fanta is about maneuvering around laws that most people recognize as unreasonable.

To get the three-day break from surveillance he needed to build his house, Mr. Irakoze paid the abanyerondo and abadasso community police units to look the other way. These are the same patrol guards who would be tasked with arresting him in the streets as an umuzunguzayi. After security patrols were taken care of, Mr. Irakoze passed cash to community-level leaders and up to cell- and sector-level leaders, all of whom are responsible for either reporting or demolishing unauthorized buildings before akajagari gets out of hand.

Paying off authorities does more than allow Mr. Irakoze to build a house; it also ensures protection for that house. Once he pays off community and cell leaders, they become obligated to defend the presence of his house if city officials do find it. There are only two explanations for an unauthorized house on land that does not have an official title that is built outside of master plan codes: either someone took a bribe, or the house predates the plan. Community officials cannot admit to taking a bribe without risking jail time. And cell and sector officials cannot admit that bribery is happening under their watch without compromising the plan's brand image of Kigali as a city with zero corruption and risking their jobs (see chapter 3).

To make all this work, though, Mr. Irakoze had to have a house, and he had to work fast. The sooner he could get a structure in place, the sooner he could say it was always there. Three abafundi (skilled masons) and ten assistant laborers can build a wattle-and-daub house in three nights. If it takes any longer, the homeowner will go broke paying bribes or get caught, and the house will be demolished.

Mr. Irakoze did succeed in embedding his unit into the city's infrastructure. As of August 2023, the house still stands in a neighborhood that has grown exponentially, and it has acquired several upgrades and an additional annex that he rents out. The neighborhood where he built it is now, as Bannyahe was, an urban center with access to electricity and water.

In an earlier article on displacement in Kigali, Vincent Manirakiza and An Ansoms also observe overnight building as a strategy of getting around demolitions. But, they caution, "despite the inventiveness of these strategies, in the end, people risk losing their investment when their zone is expropriated in favour of an infrastructural development project" (2014, 193). While Mr. Irakoze has so far successfully stayed put, individual houses do get demolished all the time. And of course, so do entire neighborhoods. But Manirakiza and Ansoms never ask the next question: if it is so risky to build without a permit, why does nearly everyone—so many people that the government estimates that 63 percent of the city's surface is covered with unauthorized buildings, and the proportion is growing (MINECOFIN 2023)—take that risk?

"Because," Mr. Irakoze laughed when I asked him, "we are many [*kubera turi benshi*]!" Like the street economy, there is nothing minor about building cowboy. It is a numbers game that involves large numbers. Houses do get torn down, just like the streets do get raided. As Mr. Irakoze explained, "If the police raid UTC [an area downtown], we escape [the raid] and continue in Nyabugogo. If they raid Nyabugogo, we continue in UTC. They cannot take us all. It's the same [*ni kimwe*]. They demolished Kiyovu, I went to Lafreshaîre. They demolished Lafreshaîre, we rebuilt here. If they demolish us here, we rebuild over there—like that."

The "we" is important. And so is Mr. Irakoze's ability to anticipate both demolition and his future presence in the city despite the inevitable processes of destruction. Like the street economy, pushing back on the plan through illicit building and unauthorized repair works only if everyone else is participating. Because Mr. Irakoze knows akajagari is, in the words of a former mayor, "overwhelming" the efforts to evict them, he can anticipate his own future in Kigali. *If they demolish us here, we rebuild over there.* Using the technologies of earth construction, distributing little somethings to neutralize surveillance, and working at night, Mr. Irakoze will rebuild his house in three days. Together, these processes (however imperfect, messy, and not the solution to all things)—community level titling; rapid, earth/clay-based building; strategic bribery; and the street economy—forge the base to repair the city from the inevitable destruction brought on by sustainable urbanism. Whether of an individual house or an entire neighborhood, demolition—however unfortunate and costly—is never total erasure, but the beginning of more unauthorized repair.

BREAKDOWN

In October 2017, the City of Kigali announced to Bannyahe landowners that their neighborhood would be demolished because of a master plan project. To cheapen the value of the neighborhood and to justify expropriation, government ministers classified Bannyahe much in the same way that scholarship of the Rwandan state classified it: as an "unsustainable" and "informal" settlement. It was an "unplanned settlement," meaning it apparently came out of nowhere and therefore "lacked basic services and access to sanitation" (Nikuze, Sliuzas, and Flacke 2020, 4). But not only did Bannyahe have access to basic services and sanitation that had been paid for and built by the residents who were scheduled to move, it was the outcome of top-down sustainable planning—generated by the demolitions in the urban core.

And as imperfect as it was, Bannyahe was built with low-cost, low-carbon, local materials. Most landowners had their titles. And the pond filtration system and wetlands agriculture below the neighborhood aligned with the Environmental Treatment Zones envisioned in the first OZ plan. A small section of the neighborhood was in the high-risk zone near the wetlands. But the problem with Bannyahe was not that it was unplanned but that it was built and owned by the wrong people: Kigali residents, many of whom had been displaced by the master plan and who organized through collective associations to repair and maintain the city, rather than foreign investors.

The project to demolish Bannyahe was first the outcome of a Finnish private equity firm and a Rwandan real estate mogul who used the plan and the City of Kigali as an instrument to get access to land and consumer markets. Converting Bannyahe into green capital required devaluing the neighborhood and the labor that went into constructing it from the ruins of master plan demolition. The firm, in partnership with the City of Kigali, would build a $56 million "sustainable" luxury housing estate called Savannah Creek over Bannyahe. In addition, the firm would invest $11 million to build the Busanza Estate, made of several three-story buildings that housed one-bedroom apartments and were classified as affordable. This was above the $10 million mark, an incentive that would have qualified the company for a reduction in its corporate income tax rate, from 30 percent to 0 percent (see Uwayezu and de Vries 2020a). At the time, the firm boasted among its investments in Kigali a $40 million multistory shopping center on the Kigali Convention Center grounds that caters mostly to expatriates and conference-goers. It makes sense that Bannyahe was chosen as the site for Savannah Creek because (like Kiyovu cy'abakene) it already had a well-serviced infrastructural base. In addition, Bannyahe was surrounded by other elite estates.

The demolition became controversial when city authorities chose an "in-kind" compensation scheme to provide both land for the Savannah Creek Estate and consumers for the affordable apartments. In violation of the national expropriation law, and in a move that was similar to the Batsinda Estate (see chapter 4), city authorities planned to compensate Bannyahe landowners with apartments in Busanza rather than cash (Nikuze, Sliuzas, and Flacke 2020, 10).

Unsurprisingly, Bannyahe landowners contested the compensation scheme (see Nikuze, Sliuzas, and Flacke 2020, 14 for excellent documentation of the disputes). This was the context in which the minister of local government and the mayor held a meeting just outside Bannyahe on April 3, 2018. In this meeting with the mayor and minister of local governance, it was clear that the dispute was over the value that residents had sunk into their properties and their neighborhood, not a "struggle to remain." People asked if they could speak to the developer directly. Some of the landholders reminded city authorities about the expropriation law that guaranteed them the right to fair market compensation for their land and houses. According to press reports, the mayor responded by declaring that the master plan was also a law. "The truth is, the master plan is the law [*ukuri ni uko igishushanyombonera ari itegeko*—literally, 'the truth is this design is the law']!" (Bukuru 2018). Then, in a moment of frustration, the minister of local government lashed out at the meeting attendees with a threat that gave the Kinyarwanda papers their headlines: "Those who want to oppose us [the government] can go ahead," he warns, "but then we will oppose you [*abashaka guhangana bajye hariya duhangane*]!"[17]

The master plan may be a law, but there were other laws that contradicted it. When the City of Kigali refused to negotiate with Bannyahe landowners, a small group went to court (see Hudani 2024, 136–44; Nikuze, Sliuzas, and Flacke 2020, 19). In Kinyarwanda media, Bannyahe residents are on record as not resisting the move. Landowners demanded cash compensation for the value of their labor, the infrastructure they had built, and their homes so they could do what they did after the demolitions in the urban core: draw on generations of knowledge to repair their properties elsewhere instead of being coerced to consume apartments that they did not want (see, e.g., Ukwezi TV 2022, https://www.youtube.com/watch?v=d0e7jxmYue0, for excellent video documentation of a meeting between Bannyahe residents and city authorities that confirms this). The court cases dragged on. Months became years. In the meantime, Bannyahe landowners did what Mr. Mutoni said they would do—what abazunguzayi also do—they disaggregated their properties to component parts and rebuilt them elsewhere.

In 2018, on a return trip to Kigali as the court cases dragged on, I caught up with my friend who first invited me to Bannyahe. He had kept his rental properties up but had already broken down his house and reassembled it in a fast-growing neighborhood on the other side of the city. Three of his neighbors from Bannyahe had done the same within walking distance of his house. I asked him why he didn't join the court case. He looked at me sideways: "*Kurwana na reta* (Fight with the state)?"

When a small group of landowners did fight the City of Kigali in the courts, he and his neighbors took advantage of the delays to disassemble their houses to component parts. They sold what they could not move, took their roofs and poles, and repaired their access to the city in other neighborhoods as they had done in Kagugu, in Gikondo, and on Mount Kigali.

Other scholars have reported this breakdown of homes in Bannyahe as an act of "domicide" that residents carried out on their own neighborhood (Hudani 2024, 144). This is an interesting formulation and calls attention to the ways that Bannyahe residents engaged in their own acts of breakdown before the police moved in. But domicide is the annihilation of housing and infrastructure by one group of people to render an area uninhabitable by another, such as bombing a residential neighborhood. In addition, this formulation could cheapen the value of expertise and the vital labor of repair and maintenance work that went into making Bannyahe. And it was that value that Bannyahe residents argued they were being undercompensated for in the court case.

By breaking down their homes and rentals into component parts and rebuilding them in other areas, Bannyahe landowners were engaging in collective forms of unauthorized repair. They were restoring not just the built environment but their control over the popular city. Built environments can be destroyed, expropriation laws can be broken, value can be plundered. But akajagari, as the product of technologies of unauthorized repair and maintenance that the entire city depends on, remains.

POSTSCRIPT: MASTER PLAN MALFUNCTION

On September 15, 2022, five years after the City of Kigali announced the demolition of Bannyahe for a master plan project, and after most Bannyahe landowners had left, the Rwanda National Police entered the neighborhood. Their first act was to turn off water and electricity that city officials and some policy scholars claimed did not exist. Then, the police moved in on the neighborhood to evict everyone who had not already left (see Rwanda Updates 2022, https://www.youtube.com/watch?v=4huAooWMCCk, for video footage of the police marching into the neighborhood). Some have suggested that because the demolition went ahead, it was an example of a state fiat. But the goal was never demolition for demolition's sake. It was driven by the principles of market-driven sustainable urbanism. The goal was to destroy Bannyahe so that the neighborhood could be rebuilt by a Finnish hedge fund as a sustainable high-end housing estate that would achieve a synergy between the environment and market growth; and it was to generate consumer demand for Busanza Estate by eliminating Bannyahe's already existing housing stock.

There is no record of what happened to Savannah Creek or the investors. During the five years of court cases, the Finnish investors backed out and sold their other holdings in Kigali. Some former Bannyahe residents ended up in Busanza Estate. Most, however, use the one-bedroom apartments they did not want as illicit rental properties. At writing, the demolition of Bannyahe seems to have achieved only what Jacob Doherty, in his 2022 work on Kampala, calls "uncreative destruction." The city was unable to secure the investment "intended to retroactively legitimize eviction" (J. Doherty 2022, 60). In the process, the demolition also took out a large chunk of the state: the infrastructure and institutions that Bannyahe residents had installed sending people to build more akajagari elsewhere. Where Bannyahe once stood is yet another vacant field in the city that green capital has not valorized. The quick repair of Bannyahe, in Kagugu, in Gikondo, and on Mount Kigali, is also evidence of the inability of the state to match the breakneck speed with which Kigali's cowboys and abazunguzayi can hack into systems of surveillance through their access to the wetland ecosystems, the street economies, and the numbers game.

As the next chapter argues, a different type of repair, recycling—driven by a logic of working with discarded objects and spaces that are broken down into component parts and rebuilt into something new—shows how the relations of disposal are reproduced and challenged under conditions of green capitalism.

6

Recycle

Wasted Space

From October 2014 to October 2018, the action in Biryogo Market began every morning at sunrise, before it officially opened. Thousands of retailers of all types—market traders, abazunguzayi, wholesalers, and porters—pushed into every available space in the market. Shouts and insults flew along with cash and clothes. The *abashoramari*—the investor-wholesalers who profited from the trade—fed stacks of cash into electronic counters in strategically located depots as forty-five-kilogram bundles of clothes moved out of their storage spaces on the streets outside and into the market on the backs and heads of day laborers. Their clients, small-time wholesalers who moved the contents of these bundles, stood on any available platform: a table that would be used as a vendor's stall later in the day or a ledge in the wall around the market. They reached into their bundles; pulled out a shirt, trousers, or pair of jeans; gave whatever came out a good shake and held it aloft for inspection; and scanned the crowd for an acceptable offer.

The buyers bid by thrusting fingers in the air indicating how many thousands of francs they were willing to pay for an article of clothing (see figure 17). Shouting an insult like "*komanyoko, sha* [hey, motherfucker]!" helped attract notice. The wholesaler accepted a bid by tossing the item on display to the buyer, who threw back a crumpled wad of cash. The buyer then had a few seconds to inspect the article and decide whether to keep it or toss it back for a refund. What determined the price of an article—what buyers were really looking for—was not the garment's value as is but its convertibility into something else. How much work would be required to break it down and make it whole again, or perhaps make it into something entirely different?

After the morning wholesale trade, retailers walked their newly acquired stock over to the cooperatives of *abadozi* (tailors and cobblers) that occupied a

FIGURE 17. The early morning caguwa wholesale in Biryogo. Photo by Francis Habimana, 2017, used with permission.

designated zone in every market. In these workshops, men and women tailors operated step-pump sewing machines that—like the fabric they worked on—were reassembled from broken parts that had been discarded by someone far away. Surrounding the machines were stacks of jobs and piles of brand tags, patches, and shoelaces salvaged from clothing items too far gone to be resold (see figure 18).

At the tailor stations, the articles of clothing went through two stages of conversion. First, tailors would break the clothes down into raw materials. Seams were cut; old tags were stripped off; sleeves and cuffs were undone; soles were removed. The parts from these out-of-date, out-of-fashion discards were then expertly rebuilt using push-pump sewing machines into contemporary styles at accessible price points. The seams of baggy jeans were resewn, reconfiguring the pants as skinnies. Holes might be added to create a distressed look. Shirts, shoes, and jeans were fitted with new brand tags. Suits were retailored for contemporary wedding and business styles. Worn soles were replaced with rubber, itself salvaged from used car tires.

The final product is called *caguwa*, a Kinyarwanda neologism made from repurposing the Kiswahili verb *kuchagua* (to choose) into a noun. Until 2018 Kigali's Biryogo Market was the largest but by no means the only place with an early morning caguwa wholesale. A kilometer north, in Kimisagara Market, the

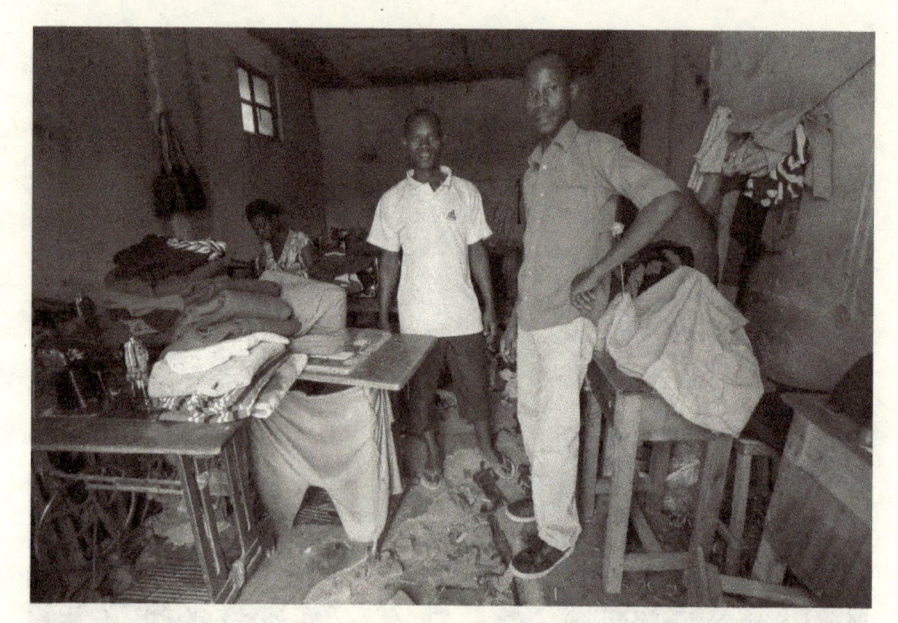

FIGURE 18. Abadozi workshop in Nyabugogo one year before it was demolished. Everything in the room, from the furniture to the sewing machines, is rebuilt from discarded parts. Photo by Bill Bamberger, 2013, used with permission.

shoe trade was (and still is at this writing) underway. Handbags were (and still are at this writing) being wholesaled down the street in Nyamirambo Market.

In addition to being a market and a center of fashion and recycling, Biryogo Market was itself a recycling project, one completed in a single day. On Thursday, October 15, 2014, Nyabugogo caguwa market—then the largest market in the city—was closed and demolished because of three competing agendas to redirect the collective value of Kigali's recycling industries to salvage other sustainable projects from failure. The result was not *in* the Kigali City Master Plan. It was made by failed efforts to generate a synergy between market growth and the environment of other areas of the city. Biryogo Market, a space that, from the perspective of a district official, had become largely unproductive, was revalorized through demolition and displacement, leaving yet another vacant lot of wasted space in the center of Kigali.

This chapter is about the recycling of discards, the recycling of space, and divergent agendas to use demolition and displacement to resolve the vacancy left behind by delayed, failed, and abandoned efforts to make Kigali sustainable. Following efforts by city authorities to salvage sustainable planning projects that fail, delay, or get abandoned not only supplies a nuanced understanding of city life, but also transforms how we understand the relationship between planning and urban authority in Kigali and elsewhere. In this chapter, as in previous chapters, I show that the destruction of built environments in Kigali is not demolition for demolition's sake but the outcome of (sometimes conflicting) agendas to convert already existing, ecologically significant processes of city building into resources for green capitalism. In the case of Nyabugogo Market, demolition had nothing to do with state control or efforts to destroy the "old" city to make the "new." It was about salvaging waste generated by sustainable urbanism.

After outlining what I mean by wasted space and recycling, I briefly sketch a recent history of the caguwa trade in Kigali and its rise to popularity after the 1994 genocide against the Tutsi. This history is important to establish why some city managers came to see the market as essential to salvaging other master plan projects from failure. I then place media made with caguwa in conversation with the Kigali City Master Plan and ecocritical scholarship because, as Karen Tranberg Hansen (2000) argues in her classic study of the global secondhand clothing trade in late twentieth-century Zambia, fashion—caguwa—is at the same time a necessity and a mode of creative self-making. This leads to an account of the multiple drafts and iterations of plans for Nyabugogo's future. As other master plan projects in the city began to go to waste, the status of Nyabugogo Market shifted from being seen by city authorities as disposable to an essential resource that could salvage other projects from failure through recycling.

I was first introduced to Kigali's caguwa trade in 2013 through a shoe seller whom I had known since 2007. Nyabugogo caguwa market was then the largest market in the city. That year, caguwa was the main source of clothing and fashion in Kigali, and it was public knowledge that there was a plan to demolish the market (see Shearer 2015). I wanted to understand how the demolition of the market would impact Nyabugogo and the rest of the city. Over several years, I got to know about a dozen retail-level caguwa sellers. Two allowed me to shadow their work in the market until it was torn down. After the market was closed, I followed the trade to Biryogo Market.

Importantly, when most of the ethnographic research for this chapter was done, the government of Rwanda had no policy on caguwa. Public discourse changed significantly in 2017, when the government of Rwanda tried (unsuccessfully) to ban secondhand imports to promote "made in Rwanda" brands. Today, the trade is heavily taxed, and government authorities invoke caguwa as a threat to Rwandan dignity, sovereignty, and the national economy. This chapter also includes ample secondary sources—media accounts and popular YouTube music videos—as evidence for claims that contradict other writing on markets in Kigali. I hope you will use the links provided to follow along with the music and visuals as you read.

WASTED SPACE AND RECYCLING

Sustainable urbanism generates a lot of waste.

Chapter 5 closed by drawing on Jacob Doherty's (2022) concept "uncreative destruction" to describe master plan malfunction in which the investment that justified the demolition of Bannyahe was eventually lost, leaving an empty field where a neighborhood once stood. In his work on Kampala, Doherty draws from Gastón Gordillo's (2014) concept of "destruction of space." For Gordillo, the concept of "creative destruction," which describes the process of making the new by destroying the old, contains a conceptual limit. In terming it "creative," Gordillo writes, "destruction is redefined as innovative, positive, desirable: the unavoidable side effect of an ever-thriving system" (2014, 80). Gordillo's critique is relevant to how urban managers in Kigali frame destruction as an inevitable process of renewal and development (see chapter 3). In his work, Doherty shows that while destruction generates much rubble and debris, these by-products also become new spaces "if only temporarily and precariously, to be captured by the displaced who... enact their own visions of urban development and futurity" (2022, 58). In Kigali, spaces like Bannyahe and the street economies they support are not outside sustainable urbanism but often emerge as unauthorized repair from master plan destruction.

For Doherty, "uncreative destruction," in which investment is delayed or never arrives, demonstrates that "waste is . . . also the constant material effect of processes of urban renewal . . . " (2022, 59; see also Gidwani 2013). In Kigali, in what Doherty calls the "spatiotemporal lags" between destruction and investment, are open fields, unoccupied buildings, and "skeleton cityscapes" (see Goodfellow 2017). Waste from sustainable urbanism shows up in popular discourse and media accounts as both *amatongo gusa* (only rubble/ruins) and, more commonly, *ubusa* (naked/vacancy). Projects that fail or are delayed also pose an issue for city managers because the utopian promise of the blank slate loses its potential the longer vacancy lingers without being developed into something else (see also Gidwani 2013, 9). To mitigate wasting, a considerable amount of effort is directed at converting vacancy into value in ways that often lead to more destruction and displacement.

By recycling, I mean the revalorization of discarded spaces and objects by working with what is already there. Repair, like the work that went into rebuilding popular neighborhoods from master plan destruction, restores that which is broken. Maintenance, like the collective work that women street traders do to move produce that is about to go bad in ways that care for the city, intervenes into the inevitable process of decay to prevent breakdowns (see chapter 5). Recycling is the conversion of material waste into value. As a commodity that begins with disposal, caguwa is also caught up in what Vinay Gidwani (2013) identifies as dialectics of waste and value. Kigali's caguwa trade is essential to a multibillion-dollar global secondhand clothing industry that ships textile discards from mostly wealthy centers of consumption to sites of disposal that are mostly in the Global South. Like wasted space, if discarded objects linger, they threaten capital, human life, and ecosystems (Corwin and Gidwani 2021, 3). The highly specialized labor of breaking secondhand clothing down and rebuilding it as caguwa, described in the opening vignette of this chapter, supports the city by transforming discards into clothing, which, in turn, becomes near zero-waste expressive culture through fashion. In other words, to borrow from Rosalyn Fredericks, caguwa is essential to reproducing the unequal global "relations of disposal" that capital depends on (2021, 2). At the same time, the case of Nyabugogo, and the recycling industries and expressive cultures that came out of it, shows that recycling, like repair and maintenance work, is also essential to maintaining the entire city against processes that are also driven by the relations of disposal—namely, the destruction of markets and their dispossession in the name of building a new "green" city.

THE MARKET THAT CAGUWA BUILT

"We buy secondhand, we sell caguwa," Mr. Georges, co-owner at JBP imports,[1] told me four months before Nyabugogo was demolished, nicely summing up the conversion process. We were standing in a room in Quartier Mateus, one of Kigali's oldest commercial districts. Bundles of caguwa were stacked throughout the room on wooden shipping crates. They were arranged according to three origins: the United States, Belgium, and China.

In 2014 Mr. Georges worked for one of over a hundred caguwa-importing companies in Kigali. They had two depots and imported, on average, ten maritime shipping containers of secondhand clothes every month. Each container had 540 *amabaro*, forty-five-kilogram bundles of used clothes. Used textile bundles were priced on a grading system based on condition of the clothes. Georges ordered a mix of high-grade (like new), medium, low-grade, and counterfeits to meet the variety of demand in Rwanda.

Secondhand clothing is not dumped in Kigali. Wealthy Rwandan investors like Mr. Georges import it. The most dangerous and low-paid work of removing secondhand clothing from waste streams and sorting the discards is done "upstream" by precarious immigrant workers in port cities in the United States, Europe, and Asia (see Hansen 2000, 18). Mr. Georges buys in bulk and then sells individual bundles of discarded clothes to individual wholesalers who then transport the bundles to markets in the early morning. His customers then auction individual items to retailers with expert knowledge of niche markets. The retailers then send each item through the conversion process in which tailors and cobblers break them down and rebuild them as caguwa, setting in motion citywide associations of sharing stock and savings.

As discussed in chapter 2, recycling was essential to postgenocide repair. While Kigali had a trade in secondhand clothing before the 1990s (see Sirven 1984; Haggblade 1990), the trade expanded significantly in the late twentieth century. The currency crashes in Rwanda followed by civil war and genocide coincided with late twentieth-century trade liberalization throughout the world. Former Eastern Bloc countries ended tariffs on used clothing imports in many countries, while textile factories closed in others. Between 1980 and 1995, global secondhand clothing trade increased sixfold, from $207 million in 1980 to $1.4 billion in 1995 (Hansen 1999, 347). By 1997 the net worth of this trade reached $14 billion and then jumped to $55 billion in 2023 (UN 2013; P. Smith 2023).[2]

In late twentieth-century Kigali, as the currency crashed and the postgenocide population surged, so did the volume of secondhand imports. In 1996 Rwanda imported just over 6.6 million kilograms of secondhand clothing. By 2022, despite several attempts by the government to ban secondhand imports, Rwanda imported more than 33 million kilograms of legal imports and an unknown quantity of smuggled clothing waste.[3] In the late 1990s, Kigali also became a dumping ground for cheap knockoffs from emerging manufacturing economies. What this meant for many consumers and in popular discourse is that the symbolic value of caguwa (secondhand) and magazini (new clothes) shifted. Caguwa (secondhand) came to be seen as an authentic, real identity marker that was alternative to the generic "new" but often knockoff commodities that were flooding the city. By the 2000s, caguwa was everywhere in Kigali—in the markets, in upscale boutiques, on the arms of abazunguzayi, and, of course, on dressed bodies. Caguwa was (as it still is at this writing) the largest source of clothing and fashion in Kigali.

In 2004 the City of Kigali closed Nyarugenge Market, then the largest market in the city. Caguwa traders moved from Nyarugenge to Nyabugogo Market, a barebones structure with concrete stalls and a corrugated roof held up by wooden poles on what was then the edge of the city. The buses followed, and Nyabugogo became the largest market and transport hub in the city.

By 2014, at Nyabugogo's height, you could not walk a straight line in the market. The only way to navigate the space was to zigzag through dense crowds of pedestrians, buses, and motorcycle taxis. To get inside the market, you had to push your way across Boulevard de Nyabugogo, a four-lane perennial traffic jam that utterly failed in its function to separate the market from the taxi park before the road arrived at Marato (the marathon, named after frequent police raids). In this stretch, hundreds of abazunguzayi hustled to catch shoppers with offers of discounted merchandise before they entered the market. Frequent security raids sent abazunguzayi running through the crowds of vendors.

Nyabugogo was also externally regulated and internally organized. The market was governed under an umbrella cooperative of traders that was governed by the Rwanda Cooperative Agency (RCA), a state regulatory board. It was divided into six administrative zones, each comprising about five hundred vendors, loosely organized by the type of clothes they specialized in. Every zone was under the supervision of a *chef*, an elected senior trader who made sure each merchant was up-to-date with taxes, rents, and security and cleaning fees. Built into the edges of each zone were *amaduka*, "shops," roughly four square meters each, and each stacked floor to ceiling with caguwa. Amaduka were the most expensive sites to rent, at RWF 50,000 per month (about $60 in 2014), and thus were usually split between two or three sellers. A step down from the amaduka were *ibibanza*, "plots," small stands in the zone's center built back-to-back. Each one measured about two square meters and rented for RWF 30,000 per month. For a little less, a retailer could rent floor space and dump a pile of clothes for bargain hunters to sift through. Prices were always negotiated, sometimes for several hours, but a generally understood ratio of rent to quality set price expectations for particular spaces. Everyone—men and women, representing every religious and ethnic background—participated in the caguwa trade in Nyabugogo.

Nyabugogo's open built space was especially well suited to the sale of caguwa. Hustling a living from a caguwa market was, as Mr. Imanzi, who specialized in the sneaker trade, reminded me during one of our interviews, not about selling but about telling people what to buy. You need to move through the market, recognize your base from the crowd, pull them into your area—often physically—and thrust the latest fashion in their face. Remind them that celebrity rappers like Jay Polly now wear Timberlands, and those Nikes you have? They are out, *sha*, being sold to some *umutarage* (peasant) in Marato right now at a discount.

Beyond the tailoring, cobbling, and retailing of caguwa, the trade in Nyabugogo was responsible for a real estate and banking boom that attracted finance capital from all over the world. Walk back toward the bus station, down Marato and past the dozens of Congolese barbershops, salons, cafés, and bars that serve the market's vendors, past the gyms with weights made from recycled brake discs from discarded automobiles, and you would find Mashirahame, a series of three-story buildings built in 2003 that rent premium space to American microfinance organizations; Kenyan, Nigerian, and Rwandan banks; and restaurants and bars. The banks financed the upper-level importing businesses of the caguwa trade, and the bars filled up with traders and tailors after 6:30 p.m. when the market closed. Across the street from the market is Kwa Mutangana (Mutangana's place), a massive structure built in 2008 to warehouse produce as it arrives from the provinces and the overflow of caguwa from Nyabugogo.

The capital that moved through Nyabugogo supported a city of craftspeople and caguwa dealers responsible for converting secondhand to fashion. But the methods that enabled the survival of individual vendors, tailors, and cobblers whom it supported are not apparent on the surface. Among the retailers, sales were intermittent, but bills were not. The temporality of fashion meant that stock needed to be offloaded quickly and always replenished with new trends. Mr. Imanzi and the other traders would often go days, sometimes weeks, without making a sale, and then receive a large payout. In the meantime, rent, school fees, and taxes would come due.

"*Amadeni* [credit]" was Mr. Imanzi's response when I asked him how he and his family weathered the financial and temporal risks of the caguwa trade. Everyone in the caguwa business lived on credit. Deferring payment allowed traders to manipulate time, keeping them afloat until the next sale while leveraging the market's associations. As in the street economy, caguwa traders in Nyabugogo also relied on each other and networks of trust, via ibimina, collective saving associations, to generate a somewhat predictable flow of cash.[4] In addition to ibimina, retailers worked together across the city to make the market work for them. Caguwa traders like Mr. Imanzi forged collectivities to activate citywide networks of consumers and commodities, using the market and social media to gain access to stock and value, and making it possible for them to offer products far beyond what was in their shop.

"*Niba bibaho, nzabishaka*," Mr. Imanzi liked to say. "If it exists, I will find it." Mr. Imanzi's clients could send him on WhatsApp a picture of any shoe they wanted, usually screenshots taken from online shopping sites like Amazon.com. Mr. Imanzi then activated the city's network of caguwa traders to find the exact pair, by size, style, and color, his client was seeking. When he had the shoes, he would confirm it with his client by sending a photo on WhatsApp; the client paid him via Mobile Money—a mobile banking application—and he would deliver the shoes in person the same day. The stockholder who provided the shoes received a commission.

In focusing on how Nyabugogo Market came into being, the forms of associational life that made it work, and the near zero-waste fashion that was made there, my aim is not to romanticize discards or lose site of the relations of disposal. It is to account for how people like Mr. Imanzi conceived of the market as a source of value that sustained the city through recycling discards into near zero-waste fashion: a conception that stood in contrast to the first two plans for Nyabugogo's future.

NYABUGOGO'S TWO SUSTAINABLE FUTURES

The first draft of Nyabugogo's "sustainable" future appeared in a 2010 Central Business District Sub Area Master Plan that was done by an all-foreign team of consultants from both OZ Architecture and Surbana. The Surbana and OZ plan proposed transforming the market through upgrades to pedestrian walkways that would connect the market to the nearby wetlands while preventing it from spilling over into the street (see figure 19). "The transformation of the Nyabugogo Market," the team writes:

> primarily aims to preserve its existing vibrant character and appealing allure to locals while at the same time, balance the redevelopment needs through developing selective sites into complimentary shopping complexes which could offer a larger variety of goods and services as well as international brands that are familiar to the tourists and expatriates ... through these measures, Nyabugogo Market is transformed from the current informal market activities into a unique and attractive tourist destination that retains the existing character while sensitively adding some complimentary [sic] and shopping experiences to the area. (OZ, Surbana, City of Kigali 2010, 5.27)

FIGURE 19. "The City Bazaar." Source: City of Kigali/OZ (2010) Final CBD Report.

Ample text and images in the plan boost the environment and carbon-free tourism as conservation. There are pedestrian walkways that will also presumably promote ecotourism in the wetlands. It promotes carbon-free transport and leisure. But there is no mention of how Nyabugogo's recycling industries are also essential to the city by mitigating the relations of disposal through transforming and de-wasting discards, all the while keeping the city clothed. The problem appears to be that the "current informal market activities" are in the way of creating a "unique and attractive tourist destination."

While the language of transformation does not call for eviction, it is difficult to imagine any other outcome from the illustrated future. In the image that accompanies the text, the barebones structures of Nyabugogo's market and the large numbers of merchants, retailers, wholesalers, and tailors are replaced with a browsing economy. Unlike Nyabugogo, which extends far beyond its built environment, all over the road and into the bus station, the proposed City Bazaar is removed from the street and contained in a plaza bordered by trees and buildings. The space is certainly efficient from a planning perspective. It facilitates the flow of traffic and ensures pedestrian safety. There are ample images of "tourists and expatriates," some dressed for hot weather and others bundled up in coats and scarves. But there is no plan for what to do with the caguwa trade that is landscaped out of the future.

In 2012 there was another draft of Nyabugogo's future. The English-language online journal *New Times* published an article announcing that Nyabugogo

Market would be converted into a RWF 30 billion (about $40 million in 2012) shopping complex (*New Times*, February 1, 2012). This time, Nyabugogo's future is presented as an eight-story, high-end shopping mall, styled in retro-industrial chic. No architect or firm is given credit for the image, and it is not apparent what makes it sustainable. It is an enclosed structure with glass elevators and exposed balconies that lead to faux smokestacks. According to the *New Times*, this eight-story complex was scheduled to break ground in June 2012. And then nothing happened.

In his work on similar digital images in Kinshasa, Filip de Boeck suggests that these images do not need to become real spaces to function. Rather, they work by training people to see urban space in normative ways by changing "the content and scale of what is deemed to be proper urban existence" (de Boeck 2011, 277). The image of Nyabugogo's future does convey a visual message about what the design team thinks a proper sustainable market should look like: a tourist-centric browsing economy that is removed from the street. In both the image from the plan and the later shopping mall, the existing caguwa market and the people who depend on it are treated as disposable, simply landscaped out with no explanation of where they would go. For de Boeck, these renditions allow for those scheduled to be evicted to "share [with the urban elite] the same dream of what that city should look like" (de Boeck 2011, 278). This sharing of the same dream can apparently explain why people can be moved without putting up a fight. Scholarship on aesthetics as a form of rule argues that to be effective to the extent to which people internalize and accept renditions of what the future city should look like, these images must be incorporated into a broader range of other sensory perceptions. They need to be taken up by people who view them as "their own inclinations" in a way that shuts down alternative claims to space (Ghertner 2015, 127).

In 2014 Kigali, no one needed to be convinced through digital imagery that a tourist bazaar or an eight-story shopping mall belonged in the city. Several multistory shopping complexes were already going up and were being promoted as master plan projects. Some of these projects were being built by foreign investors like the Finnish hedge fund that originally wanted the land below Bannyahe (see chapter 5). Other projects, like the convention center, were being built by the ruling party with public debt. There were also tourist spaces and shopping malls that predated the Kigali City Master Plan: Union Trade Center in the central business district, the famous Hotel des Milles Collines, the MTN center in Nyarutarama. With access to the best clothes, Mr. Imanzi, Mr. Gasana, and their colleagues who worked in Nyabugogo also visited these places. They used shopping centers in the way they were designed to be used. They dressed up, they strolled around, they looked at people, they took selfies and posted the photos on the same social media accounts that they used to advertise their stock. What these places were not good for was selling caguwa.

Mr. Imanzi, for example, was not convinced that the promotional images for Nyabugogo's future meant the trade and the built environment it depends on did not also belong in the city. "Caguwa," he explained when I asked him if he thought if either the tourist bazaar or the shopping mall would displace the market, "will stay because it's something that everyone loves . . . it supports a lot of people . . . and everyone wears it." Tourist bazaars and shopping malls, he reasoned, belonged in the city. But so did markets like Nyabugogo.

"Selling in the market," Mr. Imanzi explained, "is necessary . . . you sell a lot of things [*gucuruza mwisoko. Ucuruza ibintu byinshi*] . . . the profits are small [*inyungu nkeye*] . . . but because the market invites a lot of people—a lot of *different* people . . . the money keeps coming" (recorded interview, July 6, 2014). In this response, the market was a source of value for those involved in the labor of converting discards into clothing. But it was also a source of value for consumers who saw that market as more than a place for essential commodities, as a place to engage in the high-stakes labor of transforming clothing into fashion through the work of consumption ("people love it"). In other words, from the perspective of Mr. Imanzi, the images for Nyabugogo's future did not generate the uptake necessary to make a shared vision of what the city's future should look like because the designers miscalculated how valuable recycling and the fashion that went on in Nyabugogo were to the entire city.

Recycling is always about more than survival. As Karen Tranberg Hansen puts it, when understood as fashion "secondhand clothing not only mediates desires, but dresses bodies. Wearing clothes rather than rags gives dignity to people with few means" (2000, 3). The process of transforming secondhand to caguwa "reconstructs the West's cast-off clothing into a desirable commodity without making reference to its origin of provenance, in effect submerging its history of production" (2000, 3). It is a process that extends far beyond Nyabugogo Market and into how Kigali artists work with and image the city for each other.

THE ART OF RECYCLING

Cajatan Iheka's (2021) work on what he calls African ecomedia identifies an array of art movements in Senegal, Nigeria, and Kenya that use discards as an artistic medium. Artists who convert discards into art offer alternative visions of the future built from the detritus of environmental racism and the dumping industries. The Kenyan filmmaker Wanuri Kahui, the Beninois-Belgian artist Fabrice Monteiro, the Ghanaian artist El-Anatsui, and the Kenyan artist Cyrus Kabiru all engage in what Iheka calls "trash aesthetics," a form of artistic ecological activism that uses trash to make artistic statements about ecological degradation and waste that is unequally distributed to Africa. Iheka argues that one of the many insights from this work is how it recuperates discards in ways that call into question long histories of associating waste with Africa while also challenging the ways artists who identify as African are ignored in mainstream environmentalism.

"Look," Mr. Gasana, a shoe trader, told me two and a half months before Nyabugogo was demolished, when I asked him if caguwa's success had to do with low prices. "They [my clients] love/like caguwa because it is original, you know? It's real . . . people don't want to wear magazini (new fakes) . . . caguwa . . . it's real, it has marks [brands] that people know" (unrecorded interview, reconstructed from field notes, July 21, 2014). The inversion of value in this characterization of caguwa (made from recycled discards) as "real" and new clothes as "fake" is instructive. In Mr. Gasana's formulation, caguwa is valued because it provides consumers with clothes that are simultaneously definitive of urban life everywhere and unique to those who wear it. It is authentic (it has the brand). It is about making a place "in the less and less rarefied international field of cultural production" (Luvaas 2012, 4).

But since fashion is a nonverbal mode of communication, the substance of style cannot be accessed simply by speaking to wearers. The year Nyabugogo Market was demolished, caguwa fashion, or *imiyenda igezweho*, was mediating an ongoing conversation between the streets and the screen, persistently creating and remaking Kigali's image with near zero-waste fashion. For example, the 2014 hit "Caguwa," a collaboration between the late celebrity rapper Jay Polly[5] and the R&B trio Urban Boyz, equates the city's love for caguwa to the narrator's love for a woman. The song's lyrics are not about recycling but play with the double meaning of *caguwa*, using the word to mean "to choose" in one line and in the next likening a woman to caguwa clothing because she, like caguwa, is authentic. As Jay Polly's line in the chorus goes:

> *Ndakumenyesha ko uri caguwa!* (I have always known that you are caguwa [using *caguwa* in the Kinyarwanda sense as clothing]!)
>
> *Ndaguchaguye mu bandi barora.* (I chose you from all the others [using *caguwa* as the Kiswahili verb but with a Kinyarwanda prefix, *Ndaguchaguye*, I chose you].)
>
> *Uri caguwa, uratandukanye, wandutiye benshi caguwa!* (You are caguwa [metaphorically, using the noun *caguwa*, for clothes]! You are unique; you stand out above the others, caguwa!)
>
> *Uri nta magazini caguwa, wowe caguwa, ni wowe caguwa, caguwa!* (You are not *magazini* [new clothes/counterfeit], you are caguwa, it is you! Caguwa!)

In the video, Jay Polly and Urban Boyz wear clothes purchased in caguwa markets. Their wardrobe originated in every corner of the world: a Chicano Bulls sweatshirt, itself an ironic hijacking of the Chicago Bulls logo; Japanese T-shirts in kanji; New Era hats with Versace logos (see https://www.youtube.com/watch?v=1ocXUZ853Jc).

In "Caguwa" the musicians and dancers take the hodgepodge of international brands, cuts, and fashions that arrive in Kigali from all over the world via global waste streams and rework them into individual styles that are, like the woman they are singing to, different from all the rest. In the video, the woman is not wearing the rags of secondhand. She is dressed in a black skirt and aviator sunglasses that have gone through the conversion processes of breakdown and rebuilding. Life as it is portrayed in the music video for "Caguwa" extends beyond clever lyrics and choreography. The set of the video, which was filmed in a parking lot in Gikondo, is made from recycled parts. It is studded with refurbished luxury cars that are also caguwa, broken down and exported to East Africa and then repaired in the garages in Nyamirambo before being placed on the set.[6] Dancers turn the depot's office—itself a recycled shipping container—into a repurposed setting to profess love for caguwa (the woman and the fashion). Caguwa is a love song performed by a group of artists who, in 2014, were already famous in Kigali. Although the music video does not express an overt ecological theme, and my aim is not to impose outside meaning on it, it does share the same mode of production with better-known African ecomedia.

In his analysis of Fabrice Monteiro's acclaimed 2015 Africanfuturist photo essay *The Prophecy*, Cajatan Iheka (2021) shows how waste aesthetics can be used to both challenge images of undignified Africans that circulate on global mediascapes and call attention to the crimes of discarding and environmental degradation on the continent. Monteiro, who is also an engineer and a fashion designer, stages his stunning digital photographs in polluted spaces throughout Senegal—trashed coastlines, overflowing toxic landfills, smoggy cityscapes. Most of Monteiro's images are portraits of more-than-human but also recognizably female subjects with supernatural abilities rooted in Senegalese perspectives and beliefs (Cleveland 2024, 152). Monteiro's female subjects, like the woman in "Caguwa," are dressed in discards—from the ocean and from dumps—that have been styled into what Iheka identifies as "regal" fashion. This juxtaposition of regality with trash allows Monteiro to target the "slow violence" of the discard industries and their destruction of ecosystems across the continent while refusing to reproduce the trope of agentless, suffering Africans that floods Western mediascapes (Cleveland 2024, 152). The result, Iheka argues in his analysis, is an indictment of the global relations of disposal but also a conceptualization of an alternative future in which trash—and the disposability that makes it—is converted into art. As Iheka writes, Monteiro's work is a sort of crime photography. "Consumption," he goes on, "is the major crime on display" (2021, 44).

In Monteiro's photographs, unlike in "Caguwa," trash remains recognizably trash even when it is used to fashion the more-than-human subjectivity of his models. After all, Monteiro's goal is to call attention to that which we would rather ignore: the impact of overconsumption and throwaway cultures on the ecosystems where discards get dumped. Not wishing to distract from Monteiro's virtuosity or his important project, readings of his work, including Iheka's analysis, also caution that there is a real risk of losing sight of the origins and relations of disposal in Monteiro's images. Many of the places where Monteiro stages his photographs are not just ecological crime scenes but sites of intense political struggles over access to the city and recycling economies (see Fredericks 2021). This is one of the "blind spots" that Iheka identifies in his in-depth analysis of Monteiro's work—specifically, one of his most famous images, staged at the Mbeubeuss waste dump in Dakar, in which a goddess is fashioned through regal attire made from trash at the dump (see *Guardian* 2017, second image). For Iheka, the blind spot excludes the political struggle between waste pickers and the government over access to the site (2021, 58–59). There may, however, be more to this blind spot.

In her ethnographic work on the Mbeubeuss dump and the conflict between the recyclers who work there and the government, Rosalyn Fredericks shows that the dispute is about a state effort to evict the waste pickers and dispossess them of the collective value they have sunk into the land below with their own labor and resources. Their work is organized around the expert and very hazardous labor of converting discards into valuable things for reuse. While one could get the impression from Monteiro's photos that no one is addressing the issue of waste in Mbeubeuss, the waste pickers are not merely victims of dumping. They have, in fact, been limiting the toxicity that the rest of the city is exposed to. Treated as disposable by the state, they are nevertheless essential to the entire city and the industries that generate waste (Fredericks 2021, 5). There is no denying the virtuosity, vision, powerful political message, and importance of Monteiro's work. But in landscaping the waste-pickers out of the Mbeubeuss dump, his image omits a crucial citation—people who have been doing what he is often given credit for as innovative: waste workers who have been transforming discards into useful things for sixty years. Moreover, Monteiro also (however unintentionally) participates in the state project to evict people who engage in recycling by landscaping them out of his rendition of the future.

Unlike *The Prophecy*, "Caguwa" is not an internationally acclaimed art exhibition, and Jay Polly and Urban Boyz are not recognized on the international stage as environmental activists. "Caguwa" is a low-cost, lo-fi music video made by artists who work with what is already there to image the city for a Kinyarwanda-speaking audience. Still, like Monteiro's photographic subjects, the musicians, dancers, and models in "Caguwa" are fashioned with castoffs from other places into dignified urban subjectivity. If Monteiro's more-than-human subjects provide an alternative to the image of a suffering African subject surrounded by waste, "Caguwa" provides a visual aesthetics that is an alternative to both the globally circulating stock images of Rwandans as sufferers of genocide and the master plan images that landscape non-elite Rwandans out of the future.

The music video "Caguwa" does not hide the fact that the dancers are wearing discards from throwaway cultures. The name of the song allows viewers to remain cognizant of these origins, the unequal relations of disposal, and the vital labor that goes into the conversion process. The difference from Monteiro's work is that the discards in "Caguwa" do not remain trash. More than documenting ecological crimes and environmental racism, Jay Polly and Urban Boyz do something with those discards, disclosing the essential labor of recycling.

"Caguwa" was not singular. In 2014 an interdependent relationship between the recycling industries, fashion, and popular music in Kigali converted discards into aesthetic renderings of urban life. This relationship was often used to provoke debates about city life and modernity in the streets and online. For example, in his 2014 hit "Ni Danger," Danny Vumbi used caguwa fashion and language to draw the line between the country and the city, cosmopolitan modernity and ideas about "traditional" Rwandan values. The song tells the story of a young man from Biryogo, or as Vumbi calls it, "*muri* [in the] ghetto," who wishes to marry his girlfriend. After a few shots of areas that had been classified as akajagari and "informal" by planners in Biryogo, the music video takes viewers to a familiar scene: the *gusaba*, or bride wealth ceremony—staged to appear outside the city—in which the young man's family introduces themselves to their future in-laws and asks permission for the man to marry their daughter. That is when the shenanigans happen.

The two spokesmen for Vumbi's protagonist are dressed in caguwa fashion. They look obviously out of place as they face elders dressed in suits at the ceremony, staged as a village scene. One wears a black bucket hat, gold horn-rimmed sunglasses, a brown button-down shirt with a purple paisley print, a pink scarf around his neck, and black jeans above gray Italian designer boots (the boots get ample close-ups). (See https://www.youtube.com/watch?v=xFi3szI8_7s.) "Ni Danger" focuses on a street vernacular associated with Biryogo. The first spokesman addresses the young woman's family:

> *Uyu mujama akundana umubaby wanyu.* (This dude here, he is in love with your baby [using the English endearment *baby*].)
>
> *Twatuye muri ghetto, aha umwana na hirya si namubona twamufasha.* (We are from the ghetto [Biryogo] where I first saw your daughter, and we picked her up.)
>
> *Murabona ko turi kumyako?* (You can see now that we are cool/hip together, right?)

In the next verse, Vumbi gives the elder's response:

> *Sinzi niba ibyo uvuze ni ikinyarwanda.* (I do not know if the things you said are Kinyarwanda.)
>
> *Ahubwo katagira umuco karakika.* (However, you have lost [your] culture [nonliteral meaning "you have become uncivilized"].)
>
> *Sinzi ko ndamenya icyo ushaka!* (I don't know if I understand what you want!)
>
> *Kuyoka, kumwako, ibyo ni biki?* (To be "cool," "hip," what are these things?)
>
> *Umujama, umubaby reka sinshoyobwe we.* (A dude? A baby? C'mon! These things don't exist [literally "are not possible"].)

Musubira iwanyu, muzaze muvuga ibyo tumva. (Return to wherever you came from [your home]. Come back when you can say something we understand.)

"Ni Danger," like "Caguwa," was a hit in Rwanda after it was uploaded to YouTube. And it is often cited in debates about culture, language, and urban belonging.

Although Jay Polly and Urban Boyz's "Caguwa" and Danny Vumbi's "Ni Danger" are very different in genre and content, each video generates its aesthetics by transforming discards into visual art. These practices of imaging and soundtracking the city by working with what is already there have been documented in studies of more famous art movements in Dakar (see Grabski 2017) and in Lagos (Newell 2020). In Kigali this work is, to be sure, driven in part by scarcity and the ecological destruction wrought by global throwaway cultures. But it would be a mistake to simply dismiss this media as only the outcome of precarity or as unecological because it is made by artists who do not engage in dominant environmentalist discourses. The spaces and subjects in the videos provide a stark contrast to globally circulating images of Kigali—either as a site of extreme violence or as a destination for high-end green finance and tourism. In addition, these processes should be understood not merely as examples of inequality and disposability, but as inspirational models for living in cities that have been trashed by the wasting processes of capitalism.

The music videos "Caguwa" and "Ni Danger" are not in any way confrontational against urban authority, or overt calls for environmental justice, and the aim here is not to read more into the songs than what the artists intended. "Caguwa" is a love song, and "Ni Danger" offers a humorous take on generational misunderstandings. Indeed, the absence of subversion shows that, in 2014 Kigali, many musicians and their fans—like caguwa traders themselves—had not taken up the master plan aesthetics that treated recycled fashion as disposable. We can and should be critical of the relations of disposal that unequally distribute waste to cities like Kigali. But it would also be negligent to dismiss the real innovation and ecological significance of Kigali's caguwa industry as only a creative response to poverty. Ultimately, it was the cumulative value of the trade that led to Nyabugogo's demolition.

WASTED SPACE

Scholarship on state building and control in Kigali observes that when sustainable planning began, ruling party officials saw markets like Nyabugogo as potential sources of "disorder" and insecurity (Goodfellow 2022, 2012–13; Hudani 2024, 114–16). But four years after the original plan to intervene in Nyabugogo was written, it had largely been forgotten. In the meantime, the demolition of the city center that was supposed to attract outside investors was not going as planned, leaving wasted space throughout the city. The delay in attracting financing generated many other ideas about what to do with the city's markets.

By 2014 the productivity of Nyabugogo Market, operating at full capacity and converting tens of thousands of kilograms of discards into fashion every year, stood in marked contrast to the empty buildings and vacant lots generated by official efforts to make Kigali "sustainable." Six years after the destruction in the urban core began, Kiyovu cy'abakene was still mostly vacant (see chapter 4; Goodfellow 2014). Kimicanga, demolished in two stretches from 2011 to 2012, had also failed to attract investors, leaving empty fields in the middle of the city. Rather than advertising an African city rising from a postconflict tabula rasa, the vacant space was going to waste, becoming evidence of financing delays and planning failures.

As a response to the shortage of investors, the Rwanda Ministry of Trade and Industry experimented with organizing Kigali residents who were not investors themselves into cooperatives that would then become investment organizations. In this scheme, local trade cooperatives would stand as investors. They would take out loans and build projects that were in line with the Kigali City Master Plan.

In 2012 an early experiment with this type of investment began when Duhahirane-Gisozi, then a cooperative of 321 former street traders, took out a RWF 1.117 billion (approximately $1.6 million) loan at 15 percent interest to build a two-story shopping complex. The project was hailed by the government of Rwanda because, in addition to leading to new buildings, it was aimed at improving the lives of street traders through government support rather than policing and extralegal detainment (see *Rirahira*, January 11, 2012).

The Duhahirane-Gisozi shopping complex was in line with all the building requirements in the master plan. The cooperative commissioned a study that indicated a scarcity of retail space in Gisozi, a neighborhood in Gasabo District. A site was selected that had been zoned for retail. The retail complex was built within the multistory height requirements. All the required infrastructure—access to tarmac road, ample parking space—was installed. The final product, the Duhahirane-Gisozi shopping complex, was like the City Bazaar in figure 19. It was organized around a plaza that separated pedestrian traffic from the road. The building went up following all the guidelines governing the rational organization of space in ways that protected pedestrians from automobile traffic and that were easy to navigate. But it generated another temporal lag—the lag between construction and the rents that were needed to pay back the billions of francs that were advanced to build the center. Two years after the project was launched, the building was heading toward foreclosure because it lacked tenants.

In a June 8, 2014, press interview, the vice president of the Duhahirane cooperative told the Kinyarwanda online daily *Igihe* that the cooperative was only able to rent 38 of the 184 units and had to discount the units it did rent to one-third the original asking price (Twizeyimana, June 8, 2014). In the interview, the vice president tells the journalist they were being outcompeted by spaces like Nyabugogo, where merchants like Mr. Imanzi preferred to operate in powerful industries. It would be "helpful," he tells *Igihe*, if those businesses could be moved to their building. How, the *Igihe* journalist asks the cooperative vice president, would this be done? "The state," the vice president of the cooperative explained by way of an answer, "is where we find more force than anywhere else" *(Leta niyo dutezeho imbaraga nyinshi kuruta ahandi).*

In the same article, *Igihe* interviews an official at the Ministry of Trade and Industry who oversaw the Duhahirane project. The journalist asks if the empty shopping complex was due to new construction outpacing demand for high-end retail spaces. The official responds: "I do not deny," he says, "that this may be a problem . . . " The ministry official goes on: "However, this should not slow us down. [The construction] should continue because it gives confidence to the market." The official then offers a solution to the problem of vacancy. "Most of the [competing] places that are successful are not in accordance with the Master Plan . . . they want to move people to buildings that are in the Master Plan" (Twizeyimana, June 8, 2014).

In the above response, the ministry official does not clarify who "they" are. But he does acknowledge the problems created by master plan implementation. Construction must continue to maintain investor confidence and eliminate the problem of vacant lots in the city left behind by the destruction in the urban core. The rapid construction of buildings like the Duhahirane shopping complex was generating an overproduction of retail space with no tenants. In other words, the first "solution" to wasted space—using cooperatives to build master plan projects— was leading to vacant buildings that were going into foreclosure. Both the ministry official and the cooperative vice president agreed with Mr. Imanzi that no trader would voluntarily move from markets like Nyabugogo to empty shopping complexes. Duhahirane, built with all the "best practices" of rational design and planning, was being outcompeted by a giant market whose built environment consisted of a corrugated roof held up by a few posts.

In the interview, the ministry official cites the Kigali City Master Plan, not as a blueprint that needs to be followed but as a tool that can be used to justify the elimination of existing markets that were outcompeting new shopping complexes. In addition, the *Igihe* interviews show that by 2014, some authorities had come to conceive of spaces like Nyabugogo Market in ways that were very different from how they are represented in official planning documents. In the first and second plan for Nyabugogo, the market and the people who depend on it are treated as disposable. They are simply erased from the future. But by 2014, both the ministry official and the cooperative vice president in the interview above see the city's markets not as a problem that requires intervention but as a potential solution that could convert wasted space into value through recycling.

Then, on September 11, 2014, a month before Nyabugogo was torn down, there was a third plan for Nyabugogo's future that combined the 2012 shopping mall idea with the cooperative idea and also claimed to be part of the Kigali City Master Plan. The head of the caguwa cooperative Cocomanya announced to the

Kinyarwanda media that they would themselves tear down the market and rebuild it as a high-end shopping complex. In the words of the cooperative president, the new shopping mall would be "bigger and better and in line with the Master Plan" (*Kigali Today*, September 11, 2014). Again, this plan is not actually in the master plan. But in invoking the plan, the cooperative president made it a legitimate project of sustainable urbanism.

In the Cocomanya plan, the traders in the market are also seen as essential to resolving the temporal lag between demolition and investment. They will resolve the lag by becoming an investment cooperative that immediately begins construction the day after the demolition of their existing market. In the Kinyarwanda media coverage of the Cocomanya project, there is no mention that three months earlier, the Kinyarwanda media reported that the Duhahirane building, which followed the same process, was already going into foreclosure.

A month before Nyabugogo was torn down in the name of sustainable urbanism, there were several different projects that all invoked the master plan to achieve divergent ends. While every plan for the market claimed to be "in line" with the Kigali City Master Plan, there was no coherent guideline of what that actually meant or what role caguwa or the people who depended on the trade would play in the future. This situation stood in marked contrast to other scholarship on Kigali that claims market closures are examples of well-designed agendas to "securitize" urban space by destroying the "old" city. In her work, Hudani claims that "the old market . . . with its central location and buzzing activity has been seen as an unregulated space of informality, even disorder. . . . Plans to renew and modernize the markets that provided sustenance and livelihood to populations in older parts of Kigali . . . thus have affected spaces of trading as well as quartiers of habitation" (2024, 113). But this argument is extrapolated from a single government report without accounting for the actual events and contradictory aims that led to the closure and demolition of "old" markets like Nyabugogo or the time lag between the original vision and the final implementation. By the time city authorities finally closed Nyabugogo, one enterprising official had taken advantage of the delays in implementing the plan. He hijacked the original goal to "modernize" the market by converting it into a tourist bazaar and used the caguwa trade to recycle wasted space into a source of rents.

THE RECYCLING OF SPACE

On October 15, 2014, after more than four years of delayed plans to close Nyabugogo and several iterations of its future, the Rwanda National Police entered the market and shut it down. The following day, the Kinyarwanda online newspaper *Umuseke* ran an article titled "Nyabugogo: No One Knew Why the Traders Were Suddenly Moved" (*Umuseke* 2014).[7] The title was somewhat misleading as everyone *Umuseke* interviewed knew exactly why they were moved and said so. As the dust began to settle, an individual who was not involved in any of the plans for Nyabugogo's future came forward and offered to transport everyone and their stock to Biryogo Market—a market originally built in the 1990s in the same open style of Nyabugogo that predated and was certainly not "in line" with the Kigali City Master Plan.[8]

That year, Biryogo Market had nine hundred stalls that were being used by six women who were all from the neighborhood. They sold produce and *ibitenge* (printed cloth). According to one of the women who worked there and had lived in the neighborhood all her life, Biryogo Market was largely abandoned in the early 2000s. She blamed the lack of business on the 2003 demolition of Nyarugenge Market, which shifted the center of commerce several kilometers north to Nyabugogo (recorded interview, October 23, 2014).

Umuseke's reporting backed up how everyone in the market understood the move. The man who transported the stock and trade to Biryogo was a contractor in charge of collecting market taxes, security, and cleaning fees for all markets in Nyarugenge District. According to *Umuseke*, he was concerned that if the Cocomanya cooperative closed the market, the Ministry of Trade and Industry would follow through with their plan to use the closure to move traders to the Duhahirane building. Duhahirane was in another district (see *Umuseke* 2014). Nyarugenge, the district that his firm collected market taxes and fees for, would be stuck with two empty markets—Biryogo and Nyabugogo. They would be out millions of francs in monthly taxes and market fees. In other words, district-level authorities used the plan to close Nyabugogo Market to prevent other plans from taking shape so they could keep the caguwa trade in the city center.

Converting Biryogo Market into the new center of caguwa fashion was a lot like the process of making caguwa itself. The near empty Biryogo Market was, like Nyabugogo, full of rows of open stalls with shops built behind them. But the market was built for nine hundred traders. Mr. Gasana, Mr. Imanzi, and their colleagues, the tailors and cobblers, numbered three thousand. Then there were the bars, restaurants, and barbershops that depended on the caguwa trade. Caguwa tailors, cobblers, and retailers solved the problem by working with what was there. Sewing cooperatives took over surrounding rental properties; retailers took over shop space adjacent to the market.

The caguwa trade took over the streetscape outside the market, converting the area into a de facto car-free zone (see figure 20). In the process, caguwa entrepreneurs extended Biryogo Market out several city blocks beyond the limits of the original built environment and transformed the neighborhood without destroying anything or building anything new. And they did it in one day.

District-level authorities and caguwa retailers used the principles of recycling to convert wasted space into a new center of popular fashions and cultures. In the meantime, efforts to convert waste generated by master plan destruction only led to more emptiness. The plan to build a "bigger, better" shopping mall in line with

FIGURE 20. The street taken over by Biryogo Market. Photo: Francis Habimana, 2017, used with permission.

FIGURE 21. The waste of sustainable urbanism. Nyabugogo Market in 2014 (left) and 2017. Sources: photo by Claude Kobe (left), used with permission; photo (right) by Samuel Shearer.

the master plan never came into being. At this writing (eleven years later), the area where Nyabugogo Market once stood is an empty field. Without the caguwa fashion that once thrived there, it is, as the Kinyarwanda online newspaper *Kigali Today* poetically put it in 2016, "*ubusa*" (naked/vacant) (Kamuzinzi 2016) (see figure 21).

THE RE-RECYCLING OF SPACE

Biryogo remained Kigali's center of recycling and street fashion for four years. In October 2018, the City of Kigali finally closed Biryogo Market (along with several auto-part recycling operations) to remove the recycling industries from the area and to make way for a car-free tourist zone. As in the demolition of Nyabugogo, no one marched in the streets in protest. Instead, caguwa retailers and auto-part recyclers recycled more space, working with what was there in Kimisagara Market, at Kwa Mutangana (in Nyabugogo), and in Nyamirambo Market. Perhaps more significantly, however, is what a group of Kinyatrap (Kinyarwanda trap) artists did with Biryogo's wasted space after the market closed.

The story of Bushali, B-Threy, and Slum Drip's 2018 viral music video "Nituebue/Nitwebwe" (vernacular: "it's us"), filmed in the ruins of Biryogo Market, is remarkable in its own right. Before "Nituebue/Nitwebwe," all three artists were struggling to get their music on Kigali's airwaves and make ends meet in the city. "We don't come," Bushali told the Kinyarwanda vlog *The Choice Live* in a 2019 interview, "from nice places" (*The Choice Live* 2019). Just days after the demolition of Biryogo Market began, the trio occupied the rubble that was left behind and converted it into a set for a music video.

The lyrics to "Nitwebwe" are not about recycling or the environment. The song is about the trio's arrival on Kigali's hip-hop scene. The video production process and visual aesthetics, however, do make ample use of recycling. In the video, the artists occupy the rubble of Biryogo Market and repopulate it with the very subjects (working-class residents who are not "from nice places") and the recycling industries that were supposed to be evicted through demolition. The second shot of the opening scene of "Nitwebwe" is a framed view of color-coordinated caguwa sneaker fashion: Chuck Taylor limited editions, Reeboks, and Nikes. Each pair of shoes, once used up and discarded, has been broken down to its component parts and immaculately rebuilt in like-new condition. The ruins of Biryogo Market are in the background (see https://www.youtube.com/watch?v=H2w75RNd97U).

"Nitwebwe," like "Caguwa," is not singular. Bushali now has a corpus of music videos that engage with the recycling industries while imaging the city and its management in ways that are very different from the green aesthetics of market-driven sustainable urbanism. His 2018 hit "Yarababye" (One suffers), for example, was filmed down the road from Biryogo Market in front of a gym that (like most gyms in the neighborhood) is made from recycled car parts. The video itself centers around a discarded car (see https://www.youtube.com/watch?v=uondlF2iEVY). Another Bushali hit, "Kivuruga," is shot among discards in a recycling center just outside Kigali. In "Kivuruga," unlike in Monteiro's work, the workers who manage the relations of disposal remain in the frame (see https://www.youtube.com/watch?v=EfcfkAoqOJ4).

The video "Nitwebwe" is certainly an example of Kigali residents enacting their own visions of urban belonging in the rubble left behind by planning-induced destruction. "Nitwebwe" was filmed in the spatiotemporal lag between the destruction of the market and the yet-to-be-realized reinvestment in the space by the very people and economies that were supposed to be evicted by the market closure. Their work could also be compared to other art movements across the continent that make use of what is already there, converting discards into expressive culture. Although they are not formally trained in the postmodern "detritus aesthetic," coming as they do from not "nice places" (Grabski 2017, 136), "Nitwebwe" is also open to an ecocritical analysis.

The trio do not use the language of environmentalism, nor are they recognized as environmental activists in the conventional sense of the term. Still, it is precisely because Bushali, Slum Drip, and B-Threy use everyday scenes as a backdrop to their music videos that their work powerfully discloses global and local relations of disposal. Viewers get a visual sense of the discards from other places that are everywhere in Kigali, exposing residents to the environmental injustices of living and working with waste generated by the fast fashion, automobile, and other industries. "Nitwebwe" also documents the wasted space—rubble and vacancy—generated by sustainable interventions into Kigali's built environment that are concealed in planned renditions of Kigali's green future. Perhaps more importantly, we get a sense of the Rwandan artists, waste workers, tailors, cobblers, and retailers in the recycling industries who do what market-driven sustainable urbanism cannot do. While they may not overcome the relations of disposal, Kigali's essential waste workers—the artists, tailors, cobblers, retailers, and consumers—who recycle discards mitigate the unequally distributed environmental burdens of disposability. They also inspire other artists who do have an explicit environmentalist politics.

For example, in a 2022 interview, the directors of the Kinyarwanda sci-fi musical *Neptune Frost* (discussed in chapter 5), Anisia Uzeyman and Saul Williams, recognized the importance of recycling in making their film on-site in Kigali. In the interview, Uzeyman states that the lo-fi aesthetics of *Neptune Frost* were inspired by everyday recycling and repair practices they saw around them. Every set for *Neptune Frost* was put together on location with mostly found objects: the refuge is constructed from used yellow plastic jerry cans, discarded bicycle wheels, empty beer bottles. One of the main characters—Mother Board—is a cyborg who sports a prosthetic arm made from a repurposed computer motherboard taken from an e-waste recycling camp.

In the interview, Uzeyman calls attention to the ingenuity of recycling while also criticizing the conditions that make that ingenuity necessary. She cites "zero-waste" fashion as an inspiration for *Neptune Frost*. One of the plotlines in the film is a critique of the digital technology and "cleantech" green industries that mine minerals from the region that go inside smartphones and laptops, exploiting African workers and polluting ecosystems in the process. "It's like," Uzeyman says in the interview, "you come [to Kigali] on the plane with all the stuff [textiles and e-waste] that you throw away and then go back on the same plane with the cobalt and the minerals to put in the new laptops" (Bradbury 2022). In this interview, Uzeyman identifies the genealogical origins of the film's environmentalist and political commitment to making near zero-waste art: a city held up by recycling in ways that are worthy of both critique and inspiration. As Uzeyman observes, we can and should be critical of the conditions that force people to be creative with what is already there. But Kigali residents also offer ecologically significant ways to challenge those relations with alternatives to sustainable urbanism.

CONCLUSION

Market-driven sustainable urbanism generates a lot of waste that threatens other projects if it is not transformed into value for reuse. An earlier scholarship in African urban studies treats recycling of objects and space as the frenetic, ad hoc, improvised outcome of people who are making do with what is at hand because they are poor (Quayson 2014, 243). Economic precarity is certainly part of caguwa. But as the fashion and art made with caguwa—not to mention the vast archive of art made by recycling that features in ecocritical scholarship—shows, recycling is about more than poverty. And it is crucial to the entire city. It is the systemic and highly skilled work of converting toxicity into value. It is labor that reproduces the unequal relations of disposal but is also necessary to make the music, visuals, and fashion that makes life worth living in the city.

Without a doubt, some officials in Kigali wished they could demolish markets to transform the "old" city to make way for the new (Hudani 2024, 116). But these aspirations were not the source of the actual destruction and the waste that followed. Attending to the details of Nyabugogo's caguwa market—how it came into being, the recycling industries that were necessary for postgenocide repair—does not just make for a more complicated story. It shows that demolition and displacement are sometimes the outcome of ad hoc efforts to resolve the contradictions of green capitalism when planning projects fail elsewhere in the city (see also Bissell 2011, 112–13).

When the financing that justified the demolition of the urban core was delayed, the result was wasted space that, like discards in general, does not disappear. Wasted space lingers, posing a threat to future projects until it can be converted into something else. Failures and delays meant that plans needed to be reformulated as different officials shifted gears to address the issue of empty fields and buildings, leading to more demolition and displacement like the destruction of Nyabugogo Market. In other words, some of the most disruptive, wasteful, and violent demolitions in the city were not the result of a well-designed agenda or nefarious state actors trying to erase "informality." The destruction happened because consultants, city authorities, and officials in different ministries were trying to improvise solutions to other failures of sustainable urbanism.

Kigali's planners have been successful at recycling some wasted space into value. For example, in 2015, three years after Kimicanga was demolished, it became apparent that the original plan to find an investor to build a 51,000-square-meter entertainment district would not be realized. The City of Kigali managed to convert the empty field into a $14.9 million grant from the United Nations Environment Programme to upgrade it to a wetland ecotourism park.[9] But the recycling of Biryogo Market was perhaps the most successful effort at converting wasted space into value. It was completed in one day because the contractor recognized the value of the industry and, much like Kigali's unauthorized repair and maintenance workers, took advantage of the fact that the plan is not the only state project in the city. Lower-level authorities intervened in ways that caused more delay in the Kigali City Master Plan, leading to more wasted space in the city. The importance of these alternatives to green capitalism is the subject of the epilogue that follows.

Epilogue

Urban Humanities for a Broken World

This book began with the end: a December 2013 rainstorm that overwhelmed the city's infrastructure and led to a flood that Ms. Iradukunda—a former resident of Kimicanga who was pushed into the wetland neighborhood of Kosovo by master plan destruction—described as an apocalypse. The 2013 floods, driven by unpredictable rains interacting with a new road, a lack of drainage infrastructure, and the destruction of homes and workplaces, were not the end of extreme and deadly weather events in the city. In May 2016, flooding and landslides destroyed more than 5,317 built structures in Kigali and killed more than forty people (Singirankabo and Iyamuremye 2022, 2). On December 25, 2019, four people were killed during torrential flooding that happened during what is normally a dry period in the city (Ntakiyimana, Zhang, and Twagirayezu 2022, 3736). In the first five months of 2020, seventy-five Kigali residents were killed, eighteen were injured, and 692 houses were destroyed "due to extreme rains" (Singarankabo and Iyamuremye 2022, 2).

To be sure, these extreme weather events in Kigali are at least partly the outcome of the Capitalocene—five hundred years of ecological destruction brought on by the unequal distribution of environmental burdens to the majority of the world in the service of enriching a few (Moore 2015).[1] But Kigali's teams of managers and consultants have augmented these deadly disasters by intervening in the built environment, acting in ways that convert extreme weather into full-blown disasters.

For example, before the December 2013 rains, neighborhoods like Kiyovu cy'abakene and Kimicanga were destroyed to reserve valuable real estate for investors and generate demand for homes in privately built housing estates like Batsinda. Surbana's single-use zoning codes reserved the safest areas for building—flat land on top of the plateaus above the wetlands, near service infrastructures and within close distance to the city center—for investment firms like the Finnish hedge fund that wanted to build over Bannyahe (chapter 5). Consequently, the demolitions pushed Ms. Iradukunda and her neighbors down into the wetlands, where they faced an increased risk of flooding. To make the Rwandan capital convenient for visitors, the City of Kigali prioritized construction of new roads like KN 123 (see chapter 1) over updates to its increasingly strained drainage infrastructure. These added roads guided water down through their drainage pipes in a way that crushed the clay drain at the valley floor, flooding Kosovo.[2]

If "failure" means the inability to improve the human condition, then it would be easy to list the failures of sustainable urbanism in Kigali (Scott 1998; Strauss and Waldorf 2011). The master plan's zoning codes, the new road network, the neglect of drainage infrastructure, and the demolitions and displacements have outsourced the effects of climate catastrophes on the city's poorest residents to make Kigali an investment and a tourist destination. But merely identifying the errors of a project, as though those who participate in it are waiting to be corrected, is not sufficient.

In many ways, the institutions that were created to implement the Kigali City Master Plan have been successful in achieving the original 2002 goal to use planning as a way of marketing Kigali to outside investors by transforming the city's image. Now known as the "cleanest, greenest city in Africa," Kigali's MICE district boasts a convention center, dozens of high-end hotels, and several car-free retail zones. Much of this was built over the destruction in the urban core that put city residents like Ms. Iradukunda at increased risk for environmental injustices. At this writing, it is possible to land at Kigali's international airport and travel to a hotel in the central business district, go to a conference, and take a walking tour without interacting with the city's majority "informal settlements." And what might be called the failures—the interventions that have made the city more risk prone—have also been enormously productive, leading to more contracts for renewed plans.

In 2020 the City of Kigali acknowledged that the 2007 and 2013 "sustainable" master plans had made the city flood prone, segregated, and car oriented (see https://www.youtube.com/watch?v=nDRDhz-uyAA; see also Morscher and Nothdurft 2021). Instead of abandoning the project, the government of Rwanda renewed Surbana-Jurong's contract to launch another marketing campaign and write a new master plan, called "Kigali Yacu!" (Our Kigali!). The inception of "Kigali Yacu!" demonstrates just how effective crisis-driven sustainable urbanism can be, at least for those selling it. The same consultancy effectively converted the failures of its previous plans into more profitable problems and new markets for more labor and a new brand image that could be sold in the form of yet another plan.[3]

Much of the original master plan was crafted around a marketing image of the city as a postconflict tabula rasa where foreign investors could freely experiment with new urban forms that would be a better, greener city. "Kigali Yacu!" turns on an image of a different type of crisis: a city broken by a decade and a half of planning mistakes. During the virtual launch of "Kigali Yacu!," city authorities remarked that the old plan was driven by "sudden intensification" (caused by the destruction of neighborhoods like Kiyovu cy'abakene) that put the city's ecologies and the people who live there at risk. Apparently, the new plan will be guided by "organic and flexible redevelopment" and "in situ" upgrading of preexisting neighborhoods. But these buzzwords are the same that were used in the original OZ and Surbana plans, and they are followed by a renewed call for the same projects that have ruined the city for many residents: more tax incentives for outside firms to continue to profit from cheap land without redistributing surpluses, and more foreign-built "affordable housing" like the Batsinda and Busanza estates—projects that will require the demolition of the city's existing stock of affordable housing.

While much of this book was being written, the ruling party of Rwanda, working in partnership with right-wing politicians in the United Kingdom, discovered another "crisis" that leverages the relations of disposability and histories of dumping in Africa, one that traffics human beings as though they are disposable. The original plan was to remove thousands of political asylum seekers—many of whom have fled governments with human rights records comparable to Rwanda's (Macaskill 2023)—from Europe and place them in Kigali against their will. The United Kingdom paid the government of Rwanda over $904 million before the plan was scrapped there. The project has since been picked up by the Trump administration in the United States (Reuters 2025). In Kigali, the program to sell the city to right-wing politicians who wish to solve their citizenship "crisis" by exporting it to Rwanda has been framed as an economic solution to the city's luxury real estate bubble brought on by the master plan. High-end property owners in the city hope to leverage the plan to collect rents and fill empty hotels in the city's tourist districts (Soy 2022).

But not all city building in Kigali is a result of crisis-driven urbanism and green capitalism. There are projects in Kigali that *do* address life in the aftermath of political, economic, and ecological disasters: the street tech of low-carbon imperfect architectures; the unauthorized "punk urbanism" of akajagari; and the near zero-waste street fashion and media production that feature in this book. These practices are not unique to Kigali. And they are the result of long histories of racialized, gendered, and classed inequalities. But they are often silenced or ignored in work that presents Rwandans as "silent" subjects of state authority, too traumatized to build their own city.

Anyone who takes popular life in an African city on its own terms can expect to be sternly rebuked for romanticizing the survival strategies of the urban poor. But when scholars, policymakers, and consultants dismiss Kigali's popular city as a collection of deficient "informal settlements" and "slums" with no real value, they contribute to the cheapening of lives, architectures, and labor in ways that facilitate the dispossession of land, housing, and knowledges by state authorities and outside firms.

Many scholars of Kigali, for example, make passing references to the demolitions of neighborhoods like Kiyovu cy'abakene, but ignore the presence and value of ecologically significant building technologies like amatafari ya rukarakara that were at the center of these demolitions. And they overlook the distributed value generated by discount street economies in the service of measuring to what extent Kigali residents resist the state (Goodfellow 2022, 254). But as I show in chapters 4 and 5, most Kigali residents who have been displaced by sustainable urbanism have no interest in resisting the state.

They want their stuff back. They want the value of housing, infrastructures, and technologies that they built, maintained, and repaired to be restored.

This requires a different type of repair: recognition and documentation of the ecological and economic value of the technologies, infrastructures, and rental markets that were plundered from neighborhoods like Kiyovu cy'abakene and then later erased in scholarship on Kigali. That is why the ethnographic project of this book has been to show that popular Kigali is more than a case study of melancholic memory, state violence, and state building. It has been to make thinkable as viable, African alternatives to sustainable urbanism the lives, ideas, and technologies that have been silenced by finance-driven sustainable urbanism and much scholarship on Rwanda. It is to show that Kigali residents rebuilt their own city after the 1994 genocide against the Tutsi and before the ruling party hired outside firms who claimed to be repairing the Rwandan capital. And it has been to track the essential labor and knowledge that is undervalued through destruction and displacement. This is not about celebrating the heroic resistance of an urban subaltern class. It is about ethnographically showing who pays the costs for utopian efforts to find capitalist solutions to climate change and the dispossession of collective value that green capitalism depends on.

URBAN AUTHORITY

The previous chapters also seek to intervene in conventional understandings of the relationship between planning, authority, and the state in Kigali. While it would be difficult to exaggerate the influence and power of the presidency both within Rwanda and in the global imagination, few people in Kigali ever encounter upper-level ruling party authorities. The state institutions that do feature in this book are more flexible and often put together and financed by Kigali residents themselves: community police units, lower-level authorities, and institutions like umuganda. These institutions predate the RPF and the Kigali City Master Plan and do not fit neatly into the models of "high modernity" that feature in so much work on twenty-first-century Rwanda. As I show in chapters 3 and 5, the state and the extrastate networks established by international consultancies, as they actually exist for most Kigali residents, can easily be hacked into and hijacked to delay evictions while rendering building codes and laws unenforceable.

Understanding the fractured, diffuse, and heterogeneous character of urban authority also clarifies the processes of destruction and dispossession. As I have shown in chapters 3, 4, and 5, some of the most controversial demolitions in the city were not about state domination or efforts to make the city legible. At the core of these demolitions were consultants who were trying to solve problems that did not exist and far-flung efforts to match the built environment of Kigali to the fictitious representations of the Rwandan capital in planning and marketing materials. Other demolitions, like the market demolitions in chapter 6, happened through improvised efforts to address past failures that had generated wasted space in the city. Indeed, the most costly demolitions in the city were not the outcome of nefarious plots to control space. They happened as a result of often inchoate agendas that involved foreign consultants and city authorities trying to achieve outcomes that they simply could not bring into the world. These tensions between the destruction brought on by green capitalism and the popular city have relevance far beyond Kigali.

...

It is no longer controversial to claim that we are hurtling toward an apocalypse driven by several centuries of capital accumulation that has systemically cheapened nature, money, work, care, food, energy, and lives (Patel and Moore 2018). At the same time, as Nikhil Anand observes in his work on Mumbai, echoing Ms. Iradukunda, "the apocalypse is already a mundane part of everyday life for millions of people around the world" (2017, 227; see also Hecht 2023; Jackson 2014). In Kigali, city builders who work with air-dried bricks; artists and traders who recycle automobile parts and textiles that arrive in the city from other economies into fashions and creative media; and entrepreneurs who run discount street economies all engage in what Cajatan Iheka calls "infinite resourcefulness ... in a time of finite resources" (2021, 224). We can and should be critical of the conditions that force Kigali residents to recycle discards and work with what is already there. But Kigali is not, as some have suggested, "late" to urbanism. Kigali is ahead and already responding to life in a world of unpredictable weather events and limited resources in which people must deal with the discards produced by overconsumption in other places (see also Corwin and Gidwani 2021; Hecht 2023; Nixon 2011; Vergès 2017 on other contexts in the Global South). This book, therefore, is not only about Kigali. It is about what urban life might be like in the future everywhere.

My understanding of the alternatives to sustainable urbanism in this book is influenced by Cajatan Iheka's concept of "imperfect media." In his work, Iheka (2021) documents the various forms of low-fi, low-carbon, low-budget popular media that exist across the African continent, from Nollywood to South Africa's low-budget recording studios. Not only is the imperfect media production process low carbon, based on working with what is already there, but the sounds and images from imperfect media also circulate through infrastructures of recycled technologies. Iheka positions this imperfect media against the "perfect media" of Hollywood, which is "beautiful, promotes resource depletion and leaves a significant carbon footprint" (2021, 226). Imperfect media, generated from recycled tech and motivated by an aesthetic commitment to minimalism, requires different listening and viewing practices than those required by the perfect media of corporate music and film. Like Bushali's work (discussed in chapter 6), they remind us that actual urban maintenance cannot be achieved through mere conservation but requires the sometimes hazardous and messy work of converting discards into both useful and aesthetic things. Consumers of African ecomedia must learn to listen to sounds and view images more carefully, through the static and fuzz of imperfection (see also Larkin 2008). Similarly, Kigali's popular city, with its imperfect architectures, its low-budget and low-fi popular cultures of recycling, and its street economies, requires close viewing and listening practices that are not available in official policy discourse (see also Kimari 2022; Mutongi 2017; and Kinyanjui 2014 for similar arguments).

Because the chapters in this book draw ideas from select African sci-fi texts and visual art as a way of thinking differently about the future from Kigali, I conclude by suggesting a few provisional ways that Africanfuturism, or the speculative arts from Africa, could contribute to an urban humanities that seeks to tackle the reorganization of urban space on a broken planet. Africanfuturism (1) offers insight into the dys/utopias of green capitalism that leverage the climate crisis as a site of production; (2) makes the unthinkable thinkable; and (3) engages processes of cultural production that reorient our perspective on the continent while drawing attention to low-carbon, popular alternatives to sustainable urbanism.

DYS/UTOPIA

African sci-fi can offer unique perspectives on utopian projects to remake cities. In much urban theory, the word *utopia* is treated as synonymous with *fantasy* or sometimes *hope* and deployed as a self-evident critique of plans that seem unrealistic (e.g., Watson 2014). Dystopia is what you get when the hubris of dreamers comes crashing down, leaving behind a ruined world. Literary studies offer a different and very productive theory of utopia as a genre of fiction in which utopia works precisely because (unlike fantasy) it is plausible, and dystopia is the flip side of the same story rather than the catastrophe that comes at the end. This theory, which emerged in large part from the classics—critical utopias by writers like Octavia Butler and Ursula Le Guin—is augmented in works of African speculative fiction like Tendai Huchu's "The Sale" (discussed in chapter 3).

What makes Huchu's speculative fiction useful for understanding utopia is that he engages seriously with the contemporary project to build privatized smart cities across the continent. In "The Sale," Huchu begins with the premise that the dozens of contemporary plans to build optimally controlled, privatized, green cities in Africa will succeed. He then uses that success as the raw material for the future world.

In Huchu's "The Sale," all the problems that technocrats claim they can solve with foreign direct investment do get solved. Future Harare has universal health care and no crime. Water and electricity services work without interruption. The utopia becomes dystopia not when it fails but when one subject wakes up from the dreamworld, allowing Huchu to shift the narrative position from outsider to alienated insider.

From the perspective of Mr. Munyuki, Huchu's narrator, we see how foreign direct investment allows a visitor class of corporate managers to recolonize the city. The city has reinstated pass laws so that the central business district remains under the control of foreign managers, while the population is kept docile through hormones administered by "health drones." Sexuality is governed through steroids, distributed as a means of controlling the population. Biotech is used to drug the city's African residents so that they serve their sole purpose: working for the city's public-private partnership, CorpGov.

In "The Sale," Huchu teaches us how to read utopian plans to sell African cities to the world neither as failures nor as a collective dreamworld. Rather, they are designs that further fantasies about the continent as a frontier for capital expansion and a potential solution to the next series of crises generated by capitalism. In Huchu's short story, Mr. Munyuki, a "native" resident, wakes up from the drug-induced dreamworld by recalling a distant memory of an alternative city. He remembers a conversation with his father about the city before "the great sell-out." "In many ways," Mr. Munyuki tells Huchu's readers, "the city [after the sell-out] was cleaner. It had water and electricity, but it'd lost its soul, or so his father had told him, during the great sell-out" (2012, 33–34). It is the distant memory of a tolerant, imperfect city that despite its problems also has a soul that Huchu positions against the zero-tolerance urbanism of the perfectly functioning smart city. That memory drives Mr. Munyuki's push for dis-alienation throughout the rest of the story.

Rather than dismissing the images of green cities in Africa as promotional hype or celebrating them as Afro-optimist visions, Huchu takes the project of green capitalism at face value. What would it take, he asks, to build a "stable equilibrium combining free markets, scientific rationale, and central planning . . . " (Huchu 2012, 40)? How much force would be required to make a city where capital would be secure from the disasters of its own making? What would need to be sold off to rebuild a city on finance? Who would bear the costs of producing that future? In asking these questions from the estranged position of the future, Huchu surfaces a world that (among other things) calls into question the synergistic promises of green capitalism in a time when there are calls for more—not less—rigid authoritarian planning regimes based on optimized control and zero tolerance for "disorder." These questions are relevant far beyond Kigali, as African cities are once again the laboratories for building new "sustainable" and "smart" futures that will be exported elsewhere.

Africanfuturist works like "The Sale" teach us how to surface the dystopia at the core of every utopian narrative by calling attention to who bears the cost of privatized urbanism. But Africanfuturist texts and media also make thinkable what is unthinkable, offering viable alternatives to projects that claim to have no alternative.

THE UNTHINKABLE

In his work on the incomprehensibility of climate crises, the novelist and postcolonial literary theorist Amitav Ghosh notes that "fiction that deals with climate change is almost by definition not the kind that is taken seriously by serious literary journals: the mere mention of the subject is often enough to relegate a novel or a short story to the genre of science fiction" (2016, 7). Ghosh invokes science fiction to criticize the incapacity of the novel form, steeped as it is in nineteenth- and twentieth-century realism and reflecting the Enlightenment project to predict events through probability, to deal with the climate crisis in any significant way. "It is as though in the literary imagination climate change were somehow akin to extraterrestrials or interplanetary travel" (Ghosh 2016, 7). While Ghosh wants to call attention to the limits of "serious" literature by invoking a lowbrow subgenre of popular fiction, he also makes an excellent case for reading science fiction in a time of climate crises. Science fiction makes the unthinkability of living in the end times thinkable from the estranged position of an alternative world. Africanfuturism brings the subjects and places that are often erased from that future and makes their presence thinkable.

In her analysis of Lauren Beukes's *Zoo City* (discussed in chapter 5), Jessica Dickson makes an excellent case for urban theorists to engage with postcolonial science fiction as social theory. Postcolonial sci-fi has "the ability to open up 'gaps' for theorization foreclosed by ethnographic realism. Such gaps carry the potential for new ways of thinking, writing, and theorizing the city" (Dickson 2014, 67). Africanfuturism achieves this through what Dickson calls "a kind of double estrangement" (2014, 68). In this reading, the estrangement expressed in *Neptune Frost*, "The Sale," "Behind Our Irises," *Zoo City*, and *The Prophesy* echoes and inverts the estrangement generated in the Kigali City Master Plan. As discussed in chapter 3, the master plan, like utopian literature in general, achieves cognitive estrangement by inviting its audience—in this case, foreign investors—to inhabit a future world that is different from their own: better in many ways, but still a scientifically plausible future-proofed green city. The master plan is careful to work within the limits of what its audience understands to be thinkable about Kigali, presenting the city as a postconflict tabula rasa. Through imagery, it landscapes most of the people who feature in this book—cowboy architects, abazunguzayi, recyclers—out of that future while inviting its audiences in. The Africanfuturist literature discussed in this book, to borrow again from Dickson, usurps "the tools of cognitive estrangement wielded by Western Imaginaries and redirects them" while calling into question the very processes that make it possible to erase city residents from representations of the future (Dickson 2014, 68). These texts do this by foregrounding the subjects, places, and activities that are often relegated to the background as central historical agents.

In his excellent work on former combatants who inhabit the ruins of modernist architecture in Monrovia, Danny Hoffman argues that much work that treats everyday forms of repair and recycling as agency risks exaggerating what is really "survival, and nothing more" (Hoffman 2017, 155). Hoffman's work is essential for understanding how past utopian projects get lodged in built environments that urban residents must then figure out how to inhabit. But the rigid line that Hoffman draws between what is survival and what is "more" may have its own limits.

In her response to similar dismissals of South African service protests as only "minor," Gabrielle Hecht warns against making rigid distinctions between survival and environmental politics. "On the ground," Hecht writes, "poor South Africans (and others) don't slot their lives and struggles into environmental and socioeconomic boxes. Lived experience tells them that these are inseparable" (Hecht 2023, 14).

In addition, one might ask what is so insignificant about survival that it requires the addendum "nothing more"? After all, collective survival seems to be precisely what is at stake for humanity on a planet that is becoming increasingly unlivable (see Myers 2016; Tsing 2015).

The double cognitive estrangement that Dickson identifies in *Zoo City* and that also exists in *Neptune Frost*, "The Sale," "Behind Our Irises," and *The Prophesy* encourages a unique perspective—one that brings subjects and spaces that are often rendered "minor" to the center to make life in urban Africa thinkable beyond the discrete boxes of survival and resistance while acknowledging the value of popular city building. For example, how I understand much of the unauthorized repair and recycling work that features in chapters 5 and 6 builds on perspectives on urban life in *Neptune Frost* and *Zoo City*, two works that center recyclers, hackers, and hijackers as central characters who not only inhabit the city but hold it together and transform it. These works also draw our attention to the value and ecological significance of recycling, repair work, and discounting without losing sight of the inequalities at the center of that work.

Like abazunguzayi (discussed in chapter 5), the characters in *Zoo City* and *Neptune Frost* have been criminalized by authorities (but do not see themselves as criminals). When they are moved from a subordinate position to the center of the story as main characters, these works offer a unique point of view on the agentive, highly systematized, and vital labor that keeps the city (and the world) up and running.

In Kigali, when women street traders who make discount economies are foregrounded as environmental, geographic, and city-producing subjects rather than as objects of state violence, we get a better sense of how life in every sector of the city is held together. One does not need to romanticize the dangerous, low-paid, precarious work of the street economy to recognize that Kigali—a city that depends on a low-paid workforce—would fall apart without that economy. In other words, the labor of building, repair, and recycling, while not revolutionary, is without a doubt about more than simply survival.

In a very different and more in-depth study of the relationship between Black studies, Black women's geographies, and traditional geography, Katherine McKittrick calls attention to the problems of working within received traditional concepts that have, "for the most part, incorrectly deemed black populations and their attendant geographies as 'ungeographic' and/or philosophically undeveloped" (McKittrick 2006, xiii). In her work, which also draws on speculative fiction, McKittrick identifies the difficulty of recognizing that Black geographies exist "not only because sociospatial denial, objectification, and capitalist value systems render them invisible, but also because the places and spaces of blackness are adversely shaped by the basic rules of traditional geographies" (2006, 8–9). As texts, film, and photography, the Africanfuturist works cited in this book (and many more) offer a starting point to begin to open up the analysis of the places and spaces that have been rendered "ungeographic" and have also been devalued by planners and city authorities.

In an interview about *Neptune Frost*, Saul Williams puts it this way. African sci-fi is "about putting a new gaze on how we see the [African] continent; and how we see people from the continent; and how we see cultures and mythologies from the continent" (Bradbury 2022). Williams goes on, "People bandy about terms like Afrofuturism, but really what it is—it is about losing the documentarian gaze, the miserabilist gaze, that the West has profited from in how they look at the continent." Losing the miserabilist gaze means not losing sight of harm but recognizing the critical, agentive value of the city's producing, repairing, and recycling classes.

THE ALTERNATIVES

In addition to offering innovative perspectives on the literary form of sustainable master plans and the value that resides in popular urban spaces, Africanfuturist literature and media call attention to the alternatives of sustainable city building that, however much they are inspired by survival, also point to possible ways of living outside of the corporate logics of green capitalism. Part of what makes

Iheka's analysis of African ecomedia so useful is his focus on the material processes of making it. He shows how well-known African artists do not just make art that engages with ecological themes. They also commit to low-carbon, low-waste media production processes: working with what is already there and recycling trash into beautiful things in ways that challenge the assumption that residents of African cities lack the ecological consciousness to be stewards of their own urban environments (see also Caminero-Santangelo 2014).

The tailors, market traders, and consumers who depend on caguwa and the abafundi who work with amatafari ya rukarakara do not speak the language of Western environmentalism. Kinyarwanda hip-hop artists work with recycled electronics and spaces largely out of necessity. Still, these innovations nevertheless are near zero waste, low carbon, and recognizable to artists like Uzeyman and Williams who *do* engage in recycling and lo-fi aesthetics as part of a conscious commitment to environmental justice (see chapter 6). It would be easy to dismiss these imperfect architectures, recycling, and repair, too, as only survival. But they are also the processes that generate the art, music, and dense, flexible, accessible, cosmopolitan living arrangements that make life in the city worth living.

In short, Kigali's recycling industries and popular cultures, when read with Africanfuturist texts that deal with recycling, offer a perspective that simultaneously acknowledges the damage done by centuries of ecological destruction and accounts for the innovative ways that city residents build, repair, and maintain their own city on their own terms from that destruction.

This is what the urban humanities, as a project that places ethnography in conversation with literature and creative expressive culture, can do for humanity on a broken planet: draw our attention to both the sources of ecological destruction and the unexpected, if imperfect, responses from the streets.

Finally, I hope that this book, written about anticipation and the future from a place that is so often reduced to a traumatic past, nudges the conversation about Kigali in a direction that is more relevant to the people who live there, the value that is owed to them, and the futures they have already constructed. As Ms. Iradukunda reminds us in the introduction with her lighthearted response to demolition (if you want a place, you go there), the relevant question in the postapocalypse is not only how the world came to be broken (as important as that question is), but also the question asked in every Africanfuturist text: what comes after the end?

Acknowledgments

In 2007 Dr. Marina Fernando (the legend) at City College of New York suggested I spend a summer in Kigali. At CCNY, Professor David Johnson's courses on decolonization in Africa and the Caribbean and Professor Arthur Spears's courses on Black linguistics put anthropology and global Black studies together for me. Keith Joseph first introduced me to Marx, and suddenly the world made sense. Then, at CUNY Graduate Center, I met Mark Drury, Amiel Bize, and Ahmed Sharif Ibraham, a.k.a. the A$RWG, a.k.a. best friends for life. All the good ideas in this book I stole from Mark, Amiel, and Ahmed (sorry not sorry). Also at CUNY, seminars taught by Gary Wilder, Melissa Checker, Jacqueline Brown, and Katherine Verdery laid the foundations for this project.

This project began in earnest at Duke University, where I was lucky to have an amazing dissertation committee. Charlie Piot—best advisor ever (sorry I was such a handful). I learned so much from Anne-Maria Makhulu, Gary Wilder, and Anne Allison. Also at Duke, Louise Meintjes inspired this work. And thanks to the cohort for their camaraderie and collective effervescence: Brian Smithson, Yakein Abdelmagid, Patrick Galbraith, Elana Turevon, Stephanie Friede.

As the work progressed, it benefited from conversations, critical insights, and generous support from people in the United States, Rwanda, and around the world: Alyssa Miller, Matthew Omelsky, Ainehi Edoro, Vincent Joos, Tom Schilling, Eric Hirsch, Yana Stainova, and Christopher Webb all provided feedback on early drafts and have been great friends along the way. Waseem-Ahmed Bin-Kaseem has been a great fellow writer and interlocutor over the years. Robert Blunt was an early mentor and supporter. A chance meeting with Amanda Hammar in Copenhagen turned into a series of transformative conversations in several places. Bill Bamberger was always around for great times in Durham and Kigali (thanks for letting me use one of your photographs in this book). Jennie Burnett has been a wellspring of support and inspiration in the United States and Rwanda. It was great to bounce ideas off Zoe Berman in Kigali and the United States. Philip Gourevitch was always up for lively late-night debates in Kigali and New York City. Steven Jackson let me use his words as the epigraph without knowing if the book was any good (hope you don't regret it).

The situation in Kigali is . . . tricky. I hope the following hints and partial names are enough to express my gratitude. Yves's family took me in as their own back in 2007 and continue to let me wear out my welcome. Mama Yves, there are no words for how grateful I am for your support and hospitality over the years. William, Richard, and family have been such great friends over the years. Mwarimu Thérèse taught me Kinyarwanda and so much more. Kobe, wakoze (si sawas?)! Francis: Tuzi kurwana bro! Wakoze cyane muvandimwe! Thanks, too, to Baraka, Mateso, and Isaiah. Papa J—: thanks for vouching for me in some difficult-to-access places. Annie: Thank you so much for all your support and sticking your neck out for me. This research would not have been possible without you and your commitment to supporting vibrant debate and exchange of ideas in challenging environments. Thanks to the artists who gave me permission to print their lyrics, and to the hundreds of people in Kigali whom I cannot name here but who endured me day in and day out, who put up with my nosy, absurd questions and let me into their neighborhoods, workplaces, and homes. I can only hope this book does justice to your lives and the knowledge you shared with me. I owe you a debt I cannot repay. The friendly administrators at the Rwanda Ministry of Education approved this research twice: permit no 0135/2013.

Jean Allman and Rebecca Wanzo first brought me to the Washington University in St. Louis Center for the Humanities, where most of this book was written. It's difficult to put into words how grateful I am to Jean for her support and mentorship over the years. At the center, I also benefited from the support of Stephanie Kirk and Ignacio Infante. Also Terrance Wooten, co-postdoc at the center. Thanks also to Tila Neguse and Laura Perry. Billy Acree, Shefali Chandra, Nancy Reynolds, and Andrea Friedman were so supportive and welcoming in those early years at WashU.

This book finished in WashU's Department of African and African American Studies, where I am blessed with supportive colleagues and the freedom to pursue interdisciplinary scholarship and teaching. Jonathan Fenderson read multiple drafts. Every time I wanted to quit, he took me out for tacos and then told me to get back to work. Karma Frierson, friend, co-teacher, co-survivor of the Zoom and COVID years, read many drafts over the years. Shanti Parikh, Monique Bedasse, Geoff Ward, Mungai Mutoya, Gerald Early, Timothy Parsons, Sowande Mustakeem, Rafia Zafar, Marlon Bailey, Samba Diallo, Adriene Davis, Bill Maxwell, and John Baugh have all taken on the difficult task of mentoring an early-career colleague with rough edges and stage-five impostor syndrome. I am so inspired by you and grateful for your support. Thanks to Raven Maragh-Lloyd, Zachary Manditch-Prottas, Themba Mbatha, Robin McDowell, John Mundell, and Kelly Schmidt for being great friends and colleagues on this journey. I am grateful to all the students in AFAS who inspire me every day with their open-minded curiosity, critical engagement, and unwavering commitment to the project of universal liberation (easy, right?). Thanks to Shelly Mitchum and Gabrielle Weeden, who continue to do the vital work of keeping the department together against all odds. Thanks to those dudes in a St. Louis basement who saved my life.

Cajatan Iheka and Kenda Mutongi participated in an all-day Center for the Humanities first book workshop and provided invaluable critical feedback on early chapter drafts, setting it in the right direction. Danny Hoffman also provided valuable feedback on an early draft of the introduction. Thanks to Kevin O'Neill for believing in this project.

At University of California Press, Kate Marshall and Chad Attenborough took the time to edit my manuscript into a book with grace and patience. MaryAnne Hamilton made this book readable. I am also very grateful to Garth Myers and William Bissell for their service to the profession and for taking the time to review this manuscript. They made it so much better with their brilliance, which I shamelessly incorporated into the final draft as my own. Molly Webb and Saroj Thapa did an excellent job with the maps.

The research was funded by generous support from Fulbright IIE, Social Science Research Council IDRF, and National Science Foundation DDIG. The writing was funded in part by a National Endowment for the Humanities Faculty Fellowship; Washington University Center for the Humanities Faculty Fellowship; and Washington University Center for the Study of Race, Ethnicity, and Equity. Publishing the book was supported by a University of California First Gen Fellowship. An earlier version of chapter 4 appeared in the journal *City and Society*. I am grateful to Wiley for granting me permission to use it. Any errors in this book are mine. I look forward to your corrections, comrades.

Finally, I am grateful to my family for their love, support, and patience. Without you, none of this would be possible or even imaginable: kuri Mignonne, my brilliant and beautiful partner, umutima wanjye, umwamikazi wanjye. And for our daughter, Shaya, the future who joined us during the final draft with so much light and so many smiles at the time we needed her most. Mignonne and Shaya: you are the reason I wake up every morning and do this thing.

NOTES

PREFACE

1. Minibus taxis were outlawed in the city of Kigali the following year.
2. A surfboard wax that turns into a slick goo when applied to concrete ledges.
3. The skateboarding equivalent of a wheelie. Pop an ollie up a curb and land on the back wheels only, with the nose of the board up at about a thirty-degree angle.
4. Ephraim is a pseudonym used for security reasons.

1. INTRODUCTION: CAPITAL OF THE AFRICAN CENTURY

1. Ms. Iradukunda and Mr. Nsengimana are pseudonyms.
2. This account is backed up by Kinyarwanda newspaper reports of the flood that include interviews with other Kosovo residents (see *Kundabose*, December 13, 2013).
3. I follow the landscape architect Keller Easterling (2014), who coined the term, in defining *extrastatecraft* as infrastructures of global governance that both enroll and extend beyond national states.
4. The 1994 genocide against the Tutsi is the official, but contested, term. The debate around this terminology is in the phrasing *genocide against the Tutsi*, which some critics argue erases the memory of the many people who were not Tutsi who were murdered during the genocide. For a thorough review of this debate, see N. Fox (2021).
5. As Roy (2010a) argues, "slum" and "informal settlement" are useful legal fictions precisely because these concepts have no fixed definition and are therefore irrefutable.
6. In making this claim, I am not suggesting that it would have been okay to demolish Kimicanga if it were an "informal settlement" or a "slum." My point is that in Kigali, "slums" do not exist in any empirical sense, a priori of the destruction of a neighborhood. "Slums" only come into being when they are manufactured through extraeconomic violence. This

argument is consistent with over eight decades of Black studies scholarship on the United States. See, for example, St. Clair Drake and Horace R. Cayton's (1945) classic urban ethnography of Chicago's Southside, *Black Metropolis*.

7. As the former commander of the Rwandan Patriotic Front and former chief intelligence officer for Ugandan president Yoweri Museveni, President Kagame was influential in Rwanda and in the region long before he became president.

8. Specifically, the Singapore design firm Surbana addressed this reality for Kigali's leadership in a March 2, 2012, presentation, "The City of Kigali: Towards a Centre of Urban Excellence."

9. Before any of these plans appeared, in her 2002 article "Global and World Cities: A View from Off the Map," the South African geographer Jennifer Robinson predicted this situation and warned against it.

10. Also from the March 2, 2012, Surbana presentation to the City of Kigali.

11. The most salient example of this dynamic is the New York Stock Exchange, whose managers often threaten to leave New York City for better offers—and lower taxes—across the river. They never leave (and no one believes they ever will). What the NYSE is paying in higher-than-average taxes is premium rents for the use of New York's brand, the image of a command center in the global economy.

2. PRODUCTION: MAKING AN AFRICAN METROPOLIS

1. After the 1994 genocide against the Tutsi, most of the artisanal brick kilns were shut down, and the harvesting of clay is usually done now by abafundi or their aides.

2. In their research into artisanal brickmaking in rural Rwanda, An Ansoms and Jude Murison (2013) document how the government shut down artisanal brick kilns so the kilns owned by the ruling party would monopolize the production of baked bricks.

3. A famous line from Masabo Nyagenzi's song "Kavukire" (circa 1980) about demolition and roadbuilding during this period: "Harabanje haza imihanda caterepirari iratogota isambu irayishishimura, iratatamura, amabarabara atandatu ubwo inzu yanyu isigara inegetse ku mugina... ubwo inzu yanyu isigara inegetse ku mugina." (The caterpillars came to rip up your plot [but] they are not finished, they are going to make six more roads and leave your house to the termites . . . and now your house is left to the termites.)

4. Shakirah Hudani claims that Kiyovu cy'abakene was built during colonialism and "housed laborers who worked for colonial households, and for the colonial state" (2024, 38; repeated again on p. 64 and again on p. 128). Hudani's work is brilliant. And I do not wish to detract from it, but this claim does not appear to match the historical record. In this case, the chronology is important because Kiyovu cy'abakene demonstrates the technical know-how and historical agency of the people—Rwandan, Congolese, and Burundian—who played a role in partially decolonizing the city. Correctly identifying who built the neighborhood in published materials is also important for people who live there as they continue to make claims for reparations for the value that they were dispossessed of. There are ample live sources—people who once lived in Kiyovu cy'abakene, like Mr. Gutsinda—and primary and secondary sources, including maps and aerial images that indicate Kiyovu cy'abakene was built in the late 1960s after colonialism by construction workers (see, for example, Sirven 1984, 136, which draws on aerial photos during the colonial and early independence period to document the expansion of the built environment to Rugenge, the official name of

Kiyovu cy'abakene until 2007, when it was renamed Ubumwe). In addition to this evidence, material evidence in the built environment can be accessed by simply visiting the area. Until the end of colonialism in the 1960s, the area where Kiyovu cy'abakene was built, on the north side of the city and where the European Camp was located, was an armed tollbooth set up to enforce the country/camp divide, where gendarmes would have made it impossible for anyone to cross into the European Camp from where Kiyovu would later be built. The area where Kiyovu cy'abakire and Kiyovu cy'abakene meet is, to this day, called *payage* (toll). Kiyovu cy'abakene came up in the 1960s and 1970s, and it was built by African workers, not European colonial authorities.

5. *Inzoga z'abagabo*, "beer for men," as the name implies, was originally a process that kept land in the hands of men. However, even before the government of Rwanda formalized women's land rights in 2005, inzoga z'abagabo was available to female heads of households. Many of the signatories to the example in figure 4 are women.

6. In Sirven's work, he uses what was then the official name of Kiyovu cy'abakene (Rugenge).

7. To be sure, the currency crash, the outpouring of international development aid to a military ethnonationalist dictatorship, and the mismanagement of Rwanda's economy cannot explain the 1994 genocide against the Tutsi. But the currency crash and mass unemployment did create the economic conditions for a small circle of elites to recruit unemployed young people into the militias that would later be instrumental in the genocide.

8. This account illustrates the general confusion surrounding the RPA and whether they intended to liberate the city or occupy it. The person I call Ms. Kayirebwa here was later recognized by the RPF for saving several Tutsi families from Interahamwe militias by hiding people on her property. But she was also, initially, uncertain what the RPA's advance would lead to, following rumors of more violence, and she ended up on the refugee circuit before deciding Kigali was secure.

9. It was because property was already so well documented in the city that the government was able to roll out a new land registration program from 2007 to 2011.

10. This is backed up by a 2001 UN Media project Integrated Regional Information Networks (IRIN) report that documented how local defense forces and police had "stepped up" efforts to clear the streets of hawkers and street children. The 2003 Rwanda country report by Human Rights Watch quotes several informants who reported being detained through "roundups" in Nyarugenge as early as 2000.

11. Later, the national police would move the extralegal detainments to Kwa Kabuga in Gikondo, where street traders are still detained at this writing (see Shearer 2020a).

3. BRAND: RULED BY FICTIONS

1. This is because in the nineteenth century, most of the indigenous population of the area was displaced, also under the banner of "nature conservation" (see Hickcox 2007 for an excellent review of this history).

2. In 2007 English was not yet an official language of Rwanda, and only a very small subsection of the Rwandan population spoke the language.

3. While Rwanda's CO_2 emissions have increased by a few percentage points a year since sustainable development was adopted as government policy, the nation, including its capital city, Kigali, contributes far less than 1 percent to the world's global carbon footprint.

4. I never confirmed that a goat was, in fact, killed that day, although the person I call Ms. Muhizi seconded the story. The standoff with several families from Kiyovu who refused to move is backed up by other residents and news reports from that week (see *Kalisa*, June 28, 2008).

5. This definition of C4 was given during a training session on May 28, 2012, documented in that session's PowerPoint by Surbana.

6. This was good advice. Six months earlier, a Transparency International researcher who was studying corruption in Rwanda was murdered (see Michalski 2014).

7. After his term as prime minister of the United Kingdom, Tony Blair built a second career as a high-profile consultant to African governments with image issues. His client list includes Rwandan president Paul Kagame, Togo's president Faure Gnassingbé, Guinea's former president Alpha Condé, Sierra Leone's president Julius Maada, and Maada's predecessor, Ernest Bai Koroma, as well as Ghana's and Tanzania's state leadership. Because much of Blair's consulting work appears to be related to image management and branding for individual leaders, some investigative journalists have attempted to track how he gets paid for this work (see Kedem 2019).

4. DESTRUCTION: MAKING FICTIONS REAL

1. *Kurenganurwa* (reflexive verb; the closest literal translation is "to have justice restored").

2. My translation and abridgment from the original document in Kinyarwanda. The official Rwandan Parliament Chamber of Deputies record number is 601. Signed and dated as received by the Parliament Chamber of Deputies Clerk on February 17, 2010. The dashes in place of bullet points appear in the original.

3. I am not the first to identify this contradiction. Miriam Greenberg (2015) points it out in her work on what she calls "the sustainability edge." Greenberg identifies the branding processes that happen at the level of paperwork that renders the activities of environmental justice activists in the overdeveloped world unsustainable, because environmental activism does not contribute to economic growth.

4. Engineers Without Borders began the house in 2005; 2006 was when construction on the Batsinda Housing Estate began.

5. A cell executive is a leader on a scale above the umudugudu (village or neighborhood).

6. This is street vernacular, meaning not literally beaten, but defeated. It can also mean broke (in the sense of being out of money).

7. Amatafari ya rukarakara was legalized in 2019 for limited use and only for buildings that were pre-approved. Ironically, the permitting requirements and the master plan zoning codes mean that few residents in the city can work with clay within the law. Firms with the capital to go through the lengthy permitting process can use it. Thus, legalizing earth did not make every house "legal" or allow all landowners to use earth bricks, but transferred control over amatafari ya rukarakara to firms that have the financing to get approval.

5. REPAIR: PUNK URBANISM

1. Bannyahe is a moniker. The official name, which no one other than city officials uses, is Kangondo.

2. Name changed to protect the identity of the proprietor.

3. A more literal translation would be "in the place of delinquency."
4. Shakirah Hudani implies that Bannyahe predates the Kigali City Master Plan and was "formed by the various ways in which residents returned to the city after the genocide in the absence of adequate formal built space and the increased urban migration from rural areas that has accompanied rapid urban growth since that time" (Hudani 2024, 127). While a small group of farmers lived there before the master plan demolitions, the neighborhood Bannyahe itself, as part of Kigali and connected to its infrastructures, was built after the destruction began. The satellite images in figure 14 seem to provide good visual evidence that confirms the general understanding among Bannyahe residents that the neighborhood was built as a response to master plan demolitions. This chapter also respectfully differs from Hudani's claim that akajagari and neighborhoods like Bannyahe are "increasingly endangered spaces" (2024, 127). This claim does not seem to match ample secondary literature cited in this chapter and official statistics about so-called informal settlements in Kigali that cite informality as the majority of urban spaces in the city.
5. These health workers played a key role not only in mitigating the hazards of living near the wetlands but also in providing ministry-level authorities with health statistics (see Shearer 2020b).
6. This is not my original concept. See Woods (2019) for an example of scholarship that uses punk urbanism in a very different context to describe insurgent DIY city building during London's late twentieth-century punk era that focuses on music culture.
7. A play on the title of Saul Williams's (2016) Afropunk album *Martyr Loser King*, itself an act of détournement.
8. For instance, life hacking, a term used to describe minor adjustments to the self to accommodate the demands of capitalism.
9. Plural; singular, *ikimina*, a contribution to a collective savings scheme.
10. Among Kigali's street traders, though, *ibimina* enforcement is fierce, and rumors about various punishments for trying to get out of paying *ibimina* are plentiful.
11. These wages were averages in the year 2015.
12. *Abapandagari* is an insult. It comes from the Kiswahili imperative *panda gari* (get in the car).
13. Some have argued that there is an absence of protest in Kigali (see, e.g., Ukwezi TV 2016). Footage of the protest, still available on YouTube, provides a corrective to these claims.
14. Nyarugenge Market is one of the areas where there is supposed to have been a "shrinkage" in the street economy (see Goodfellow 2022). The video tells a very different story.
15. This explanation tracks with the origins of "Billism" and urban cowboys in neighboring Congo (de Boeck and Plissart 2004, 36–39).
16. Although the Kinyarwanda trap artist ZEO successfully makes a hybrid of the *umukoboyi* and the *umushumba* in his 2024 music video "Rwamakombe."
17. The phrase could also be translated more literally as "Those who want to confront us, go there [to the court], where we will confront you!"

6. RECYCLE: WASTED SPACE

1. This is a pseudonym for the company.
2. These statistics are from the UN commodity trade database and should be read as lowball estimates. The United Nations simply compiles data as it is reported by governments,

which means that used clothes that are smuggled (a significant part of the trade) are not listed. Furthermore, rags, tags, and other loose ends that are also part of the used clothes trade are not listed under the worn clothing code in the UN statistics but fall under "industrial textile rags such as wipers" and are missing from these figures. But in Kigali, as in many places, rags and tags are crucial for fixing articles before placing them on the shelf.

3. This figure includes only taxed and reported caguwa imports. There are no official statistics on what is being smuggled, but Kigali National Police and border guards publish near monthly reports of intercepting hundreds of kilograms of smuggled caguwa at a time.

4. See chapter 5 for a more detailed discussion of ibimina and how they work.

5. Jay Polly died in police custody on September 2, 2021. The official cause of death was methanol poisoning from drinking bootlegged alcohol.

6. I learned about the cars on the set through the serendipity of doing field research on caguwa while living in Nyamirambo. Many of the cars on the set of "Caguwa" were not in working condition when the video was filmed.

7. Nyabugogo: Abaturage Ntibumva impamvu bimuye huti huti; the literal translation is "the people do not understand the reason it was moved suddenly."

8. In her work, Shakirah Hudani writes: "In 2014, traders in the old Biryogo Market were relocated ahead of schedule to a new 'modernized' market area, and this disrupted their social networks and trade associations" (2024, 113). Hudani's work is excellent. But it is unlikely that this happened. I was in Biryogo and did research there all of 2014, and the market never closed. No one was moved from Biryogo Market until 2018, and they were not moved into the "modern" market in Nyarugenge. It is possible Hudani confuses Biryogo with Nyabugogo, a different market several kilometers to the northwest. In 2014 Nyabugogo Market was moved to Biryogo (although the original plan was to move people to the Duhahirane building in Gisozi). Most of the women who worked in Biryogo before the caguwa trade was moved there remained. In addition to my own ethnographic work in the market, there are ample secondary sources in English published outside of Rwanda that show that Biryogo Market was not closed in 2014. When the government announced its plan to (unsuccessfully) ban secondhand clothing, outside journalists went to Biryogo Market to interview traders. See, for example, *Al Jazeera*'s October 5, 2018, reporting on the market, with excellent images of the space demonstrating that the market had not been closed (Essa 2018). My aim is not to fact-check Hudani. But the correct details do matter in this case because they help us understand how displacement actually works, why markets were demolished, and the drivers behind displacement. The actions taken against caguwa markets were not efforts by the state to "modernize" or "securitize" the city but haphazard attempts to solve the problem of wasted space by moving the caguwa trade around.

9. See https://open.unep.org/docs/gef/CEO/5194_CEO%20Request.pdf.

EPILOGUE: URBAN HUMANITIES FOR A BROKEN WORLD

1. The entire continent of Africa generates less than 4 percent of the greenhouse gases in the atmosphere every year (Powanga and Kwaka 2024). To put this in perspective, the United States, with a population one-fourth the size of the continent of Africa, is responsible for 19 percent of the world's greenhouse gas emissions.

2. This chain of events suggests, as many scholars in the environmental humanities and urban studies have argued, that environmental justice cannot be achieved through

"technical adjustments" to the built environment, as though "climate change is a management problem that experts may resolve, rather than an ethical and moral problem that humans around the world should recognize, discuss, and address as political agents" (Günel 2019, 10). Indeed, much of the appeal of the Kigali City Master Plan to foreign investors lies in what the environmental anthropologist Melissa Checker calls the "sustainability myth"—"the inherently contradictory promise . . . that we can stimulate economic growth while mitigating the effects of climate change without any sacrifice" (2020, 7).

3. Surbana's work in Kigali led to more business not only in Kigali but also across Africa. The consulting firm used its high-profile work selling Kigali as a tourist and investment destination to grow its portfolio throughout the continent, exporting the Kigali model to the rest of Africa. The firm now has contracts to plan privatized metropolises based on the Kigali model in Kinshasa, Brazzaville, Libreville, Bujumbura, Conakry, Luanda, and Lagos, as well as secondary cities in Rwanda, Tanzania, and Ghana (see Neuwirth 2021).

REFERENCES

5280 Magazine. 2008. "Redesigning Rwanda." May 2008. https://www.5280.com/redesigning-rwanda/.

Adejunmobi, Moradewun. 2016. "Introduction: African Science Fiction." *Cambridge Journal of Postcolonial Literary Inquiry* 3 (3): 265–72.

Allison, Simon. 2017. "Like It or Not, Rwanda Is Africa's Future." *Mail & Guardian*, July 7, 2017. https://mg.co.za/article/2017-07-07-00-like-it-or-not-rwanda-is-africas-future/.

AmaG The Black and Safi Madiba. 2014. "Nyabarongo." Super Level/Jungle Entertainment Ventures, October 9. Music video, 5:00. https://www.youtube.com/watch?v=L-s9F8IIHCA.

Anand, Nikhil. 2017. *Hydraulic City: Water and the Infrastructures of Citizenship in Mumbai*. Durham, NC: Duke University Press.

Ansoms, An, and Jude Murison. 2013. "Formalizing the Informal in Rwanda: From Artisanal to Modern Brick and Tile Ovens." In *Securing Livelihoods: Informal Economy Practices and Institutions*, edited by Isabelle Hillenkamp, Frédéric Lapeyre, and Andreia Lemaître, 235–53. Oxford: Oxford University Press.

Appel, Hannah. 2019. *The Licit Life of Capitalism: U.S. Oil in Equatorial Guinea*. Durham, NC: Duke University Press.

Arefin, Mohamed Rafi, and Rosalyn Fredericks. 2025. "A Global Era of Disposability: The Anthropocene, The Apotheosis of Waste." *Antipode*. https://doi.org/10.1111/anti.13083.

Bafana, B. 2016. "Kigali Sparkles on the Hills." *Africa Renewal Online* 30:6–7. https://www.un.org/africarenewal/magazine/april-2016/kigali-sparkles-hills.

Baffoe, Gideon, Josephine Malonza, Vincent Manirakiza, and Leone Mugabe. 2020. "Understanding the Concept of Neighbourhood in Kigali, Rwanda." *Sustainability* 12 (4): 1555.

Barnes, Jessica. 2014. *Cultivating the Nile: The Everyday Politics of Water in Egypt*. Durham, NC: Duke University Press.

———. 2017. "States of Maintenance: Power, Politics, and Egypt's Irrigation Infrastructure." *Environment and Planning D: Society and Space* 35 (1): 146–64.
Becker, Sam. 2022. "Get Out of Boulder." *Boulder Weekly*, March 3, 2022. https://boulder weekly.com/news/get-out-of-boulder/.
Bedore, Pamela. 2017. *Great Utopian and Dystopian Works of Literature*. The Great Courses. Audio CD, 24 lectures.
Benjamin, Walter. (1940) 1999. *The Arcades Project*. Translated by Howard Eiland and Kevin McLaughlin. Cambridge, MA: Belknap Press.
Berger, James. 1999. *After the End: Representations of the Post-Apocalypse*. Minneapolis: University of Minnesota Press.
Beukes, Lauren. 2010. *Zoo City*. New York: Penguin.
Bhowmick, Soumya. 2019. "Toward the SDGs: The Rwanda Story." Observer Research Foundation, May 17, 2019. https://www.orfonline.org/research/toward-the-sdgs-the-rwanda-story-50935/.
Bindseil, Reinhart. 1988. *Le Rwanda et l'Allemagne depuis le temps de Richard Kandt*. Berlin: D. Reimer.
Bishumba, Nasra. 2010. "SSFR Sells Three Plots in CBD Project." *New Times*, August 17, 2010. https://www.newtimes.co.rw/article/42790/National/ssfr-sells-three-plots-in-cbd-project.
Bissell, William Cunningham. 2011. *Urban Design, Chaos, and Colonial Power in Zanzibar*. Bloomington: Indiana University Press.
Blum, Andrew. 2007. "Planning Rwanda." *Metropolis*, November 1, 2007. https://metropolis mag.com/projects/planning-rwanda/.
Botte, Roger. 1985. "Rwanda and Burundi, 1889–1930: Chronology of a Slow Assassination, Part 1." *International Journal of African Historical Studies* 18 (1): 53–91.
Bower, Jonathan, Sally Murray, Robert Buckley, and Laura Wainer. 2019. "Housing Need in Kigali." Reference no. C-38406-RWA-1.
Bradbury, Sarah. 2022. "Saul Williams and Anisia Uzeyman on *Neptune Frost*, Lo-Fi Afrofuturistic Look, Inverting Western Gaze." *The Upcoming*, November 5, 2022. Interview, 25:04. https://www.youtube.com/watch?v=RxqHSZlHmoM.
Brico, Elizabeth. 2017. "The Happiest City in the U.S. Has a Secret." *Talk Poverty*, December 12, 2017. https://talkpoverty.org/2017/12/12/happiest-city-u-s-secret/.
Brooks, Andrew. 2015. *Clothing Poverty: The Hidden World of Fast-Fashion and Secondhand Clothes*. New York: Zed Books.
Brundtland Commission. 1987. "Report of the World Commission on Environment and Development: Our Common Future."
BTN TV Rwanda. 2022. "Abazunguzayi Bakoze Igisa Nkimyigaragambyo Batesha Abanyerondo Mugenziwabo Bari Batwaye." August 29, 2022. News video, 12:44. https://www.youtube.com/watch?v=-mxaeQphtkw&list=PPSV.
———. 2023. "Kigali: Abazunguzayi Bishe Umunyerondo." March 9, 2023. News video, 12:38. https://www.youtube.com/watch?v=ekrWzfLqtgA.
Buck-Morss, Susan. 1995. "The City as Dreamworld and Catastrophe." *October* 73 (Summer): 3–26.
———. 2000. *Dreamworld and Catastrophe: The Passing of Mass Utopia in East and West*. Cambridge, MA: MIT Press.
Bukuru, J.C. 2018. "Meya Nyamulinda Yumvishije Abo Muri 'Bannyahe' Kutararikira Ingurane Y'amafaranga, Baba Ibamba." *Igihe*, January 25, 2018. https://www.igihe.com

/ubukungu/article/meya-nyamulinda-yumvishije-abo-muri-bannyahe-kutararikira-ingurane-y-amafaranga.

Burnett, Jennie. 2012. *Genocide Lives in Us: Women, Memory, and Silence in Rwanda*. Madison: University of Wisconsin Press.

Bushali. 2018. "Yarababaye." Green Ferry Music. Music video, 3:17. https://www.youtube.com/watch?v=uondlF2iEVY.

Bushali and Ghanaian Stallion. 2021. "Kivurugu." Music video, 3:06. https://www.youtube.com/watch?v=EfcfkAoqOJ4.

Bushali, Slum Drip, and B-Threy. 2018. "Nituebue." Green Ferry Music. Music video, 3:32. https://www.youtube.com/watch?v=H2w75RNd97U.

Caminero-Santangelo, Byron. 2014. *Different Shades of Green: African Literature, Environmental Justice, and Political Ecology*. Charlottesville: University of Virginia Press.

Carminero-Santangelo, Byron, and Garth Myers. 2011. *Environment at the Margins: Literary and Environmental Studies in Africa*. Athens: Ohio University Press.

Carmondy, Pádraig R., James T. Murphy, Richard Grant, and Francis Y. Owusu. 2023. *The Urban Question in Africa: Uneven Geographies of Transition*. Hoboken, NJ: Wiley.

Chakrabarty, Dipesh. 2007. *Provincializing Europe: Postcolonial Thought and Historical Difference*. Princeton, NJ: Princeton University Press.

Chalfin, Brenda. 2023. *Waste Works: Vital Politics in Urban Ghana*. Durham, NC: Duke University Press.

Chatterjee, Partha. 2004. *Politics of the Governed: Reflections on Popular Politics in Most of the World*. New York: Columbia University Press.

Chattopadhyay, Swati. 2012. *Unlearning the City: Infrastructure in a New Optical Field*. Minneapolis: University of Minnesota Press.

Checker, Melissa. 2011. "Wiped Out by the 'Greenwave': Environmental Gentrification and the Paradoxical Politics of Environmental Sustainability." *City & Society* 23 (2): 210–29.

———. 2020. *The Sustainability Myth: Environmental Gentrification and the Politics of Justice*. New York: NYU Press.

The Choice Live. 2019. "Nituebuee Cg Nimwebwe? Dukunzwe Mu Gihugu Hose: KinyaTrap Mukiganiro Kiryoshye." May 18, 2019. Interview, 32:21. https://www.youtube.com/watch?v=EZsKSCcNaOE.

Chrétien, Jean-Pierre. 2003. *The Great Lakes of Africa: Two Thousand Years of History*. Princeton, NJ: Zone Books.

Cities Alliance. 2008. "Cities without Slums Annual Report." Accessed June 3, 2018. https://www.citiesalliance.org/resources/publications/annual-reports/annual-report-2008.

City of Boulder. 2013. "Conversations with Extraordinary People: Carl Worthington." Interview, 58:22. https://vimeo.com/64429103.

City of Kigali. 2002. *Kigali Economic Development Strategy*. August. https://grandslacs.graduateinstitute.ch/files_on_st7210/2693.pdf.

———. 2020. "Launch of Master Plan." September 4. Streamed event, 3:13:49. https://www.youtube.com/watch?v=nDRDhz-uyAA.

Clark, Larry, dir. 1995. *Kids*. Written by Harmony Korine. Shining Excalibur Films, 1:31:00.

Cleveland, Kimberly. 2024. *Africanfuturism: African Imaginings of Other Times, Spaces, and Worlds*. Athens: Ohio University Press.

CNN. 2013. "Africa's New Cities: Future or Utopian Fantasies?" *CNN Business*, May 30, 2013. https://www.cnn.com/2013/05/30/business/africa-new-cities-konza-eko/index.html.

Collier, Stephan J., Jamie Cross, Peter Redfield, and Alice Street. 2017. "Little Development Devices." *Limn* 9. https://limn.press/issue/little-development-devices-humanitarian-goods/.

Coquéry-Vidrovitch, Cathérine. 2012. "Racial and Social Zoning in African Cities from Colonization to Postindependence." In *Beyond Empire and Nation: The Decolonization of African and Asian Societies, 1930s–1970s*, edited by Els Bogaerts and Remco Raben, 267–86. Leiden, Netherlands: Brill.

Cortina, Matt. 2015. "Black in Boulder." *Boulder Weekly*, March 19, 2015. https://boulderweekly.com/news/black-in-boulder/.

Corwin, Julia E., and Vinay Gidwani. 2021. "Repair Work as Care: On Maintaining the Planet in the Capitalocene." *Antipode*. https://doi.org/10.1111/anti.12791.

Crisafulli, Patricia, and Andrea Redmond. 2012. *Rwanda Inc.: How a Devastated Nation Became a Model for the Developing World*. New York: Palgrave Macmillan.

Csicsery-Ronay, Istvan. 1988. "Cyberpunk and Neuromanticism." *Mississippi Review* 16 (2/3): 266–78.

Davis, Mike. 2004. "Planet of Slums." *New Left Review* 26 (March/April): 5–34.

———. 2006. *Planet of Slums: Urban Involution and the Informal Working Class*. New York: Verso.

Death, Carl. 2015. "Four Discourses of the Green Economy in the Global South." *Third World Quarterly* 36 (12): 2207–24.

de Boeck, Filip, and Marie-Françoise Plissart. 2004. *Kinshasa: Tales of the Invisible City*. Ghent: Ludion; Tervuren: Royal Museum for Central Africa.

de Boeck, Filip. 2011. "Inhabiting Ocular Ground: Kinshasa's Future in the Light of Congo's Spectral Urban Politics." *Cultural Anthropology* 26 (2): 263–86.

de Boeck, Filip, and Sammy Baloji. 2016. *Suturing the City: Living Together in Congo's Urban Worlds*. London: Autograph.

Degani, Michael. 2022. *The City Electric: Infrastructure and Ingenuity in Postsocialist Tanzania*. Durham, NC: Duke University Press.

de la Mairieu, Baudouin Paternostre. 1983. *Le Rwanda, son effort de développement: Antécédents historiques et conquêtes de la révolution Rwandaise*. Brussels: A. de Boeck.

de Montclos, Marc-Antoine Pérouse. 2000. *Kigali après la guerre: La question foncière et l'accès au logement*. Paris: Centre Français sur la Population et le Développement.

Desrosiers, Marie-Eve. 2014. "Rethinking Political Rhetoric and Authority during Rwanda's First and Second Republics." *Africa* 84 (2): 199–225.

De Witte, Ludo. 2002. *The Assassination of Lumumba*. New York: Verso.

Dickson, Jessica. 2014. "Reading the (Zoo) City: The Social Realities and Science Fiction of Johannesburg." *Johannesburg Salon* 7:67–78.

Dijkstra, Andrea. 2020. "Rwanda's Clothing Spat with the US Helps China." *BBC*, September 30, 2020. https://www.bbc.com/news/business-54164397.

Doherty, Jacob. 2022. *Wasteworlds: Inhabiting Kampala's Infrastructures of Disposability*. Oakland, CA: University of California Press.

Doherty, Killian. 2014. "The Metamorphosis of Post-Genocide Kigali." *Failed Architecture*, May 19, 2014. https://failedarchitecture.com/the-metamorphosis-of-post-genocide-kigali/.

Doughty, Kristin. 2020. "Carceral Repair: Methane Extraction in Lake Kivu, Rwanda." *Cambridge Journal of Anthropology* 38 (2): 19–37.

Drake, St. Clair, and Horace R. Cayton. 1945. *Black Metropolis: A Study of Negro Life in a Northern City*. Chicago: University of Chicago Press.

Easterling, Keller. 2014. *Extrastatecraft: The Power of Infrastructure Space*. New York: Verso.

Eisinger, Peter. 2000. "The Politics of Bread and Circuses: Building the City for the Visitor Class." *Urban Affairs Review* 35 (3): 316–33.

Elleh, Nnamdi. 2002. *Architecture and Power in Africa*. Westport, CT: Praeger Publishers.

Elyachar, Julia. 2012. "Next Practices: Knowledge, Infrastructure, and Public Goods at the Bottom of the Pyramid." *Public Culture* 24 (66): 109–29.

Enwezor, Okwui. 2010. "Modernity and Postcolonial Ambivalence." *South Atlantic Quarterly* 109 (3): 595–620.

Esmail, Shakirah, and Jason Corburn. 2020. "Struggles to Remain in Kigali's 'Unplanned' Settlements: The Case of Bannyahe." *Environment and Urbanization* 32 (1): 19–36.

Essa, Azad. 2018. "The Politics of Second-Hand Clothes: A Debate Over 'Dignity.'" *Al Jazeera*, October 5, 2018. https://www.aljazeera.com/features/2018/10/5/the-politics-of-second-hand-clothes-a-debate-over-dignity.

Fabian, Johannes. 2000. *Out of Our Minds: Reason and Madness in the Exploration of Central Africa*. Berkeley: University of California Press.

Finn, Brandon Marc. 2018. "Quietly Chasing Kigali: Young Men and the Intolerance of Informality in Rwanda's Capital City." *Urban Forum* 29:205–18.

Fox, Justin. 2013. "Why President Kagame Runs Rwanda like a Business." *Harvard Business Review*, April 4, 2013. https://hbr.org/2013/04/why-president-kagame-runs-rwanda.

Fox, Nick J. 2023. "Green Capitalism, Climate Change and the Technological Fix: A More-than-Human Assessment." *Sociological Review* 71 (5): 1115–34.

Fox, Nicole. 2021. *After Genocide: Memory and Reconciliation in Rwanda*. Madison: University of Wisconsin Press.

Fredericks, Rosalyn. 2021. "Anthropocenic Discards: Embodied Infrastructures and Uncanny Exposures at Dakar's Dump." *Antipode*. 1–22 https://doi.org/10.1111/anti.12796.

Gakire, Anastase. 2008. "La population de Rugenge passe la nuit à la Belle Étoile après démolition de leurs maisons." Kigali: Ligue des Droits de la personne dans la région des Grands Lacs (LDGL), July 23, 2008.

Gastrow, Claudia. 2017a. "Aesthetic Dissent: Urban Redevelopment and Political Belonging in Luanda, Angola." *Antipode* 49:377–96.

———. 2017b. "Cement Citizens: Housing, Demolition and Political Belonging in Luanda, Angola." *Citizenship Studies* 21 (2): 224–39.

———. 2024. *The Aesthetics of Belonging: Indigenous Urbanism and City Building in Oil Boom Luanda*. Chapel Hill: University of North Carolina Press.

Gaugler, Jennifer. 2018. "Modern Materials for Dwelling: The Evolution of Durability and Domesticity in Rwandan Housing." *Traditional Dwellings and Settlements Review* 29 (2): 39–54.

Ghertner, D. Asher. 2015. *Rule by Aesthetics: World-Class City Making in Delhi*. New York: Oxford University Press.

Ghosh, Amitav. 2016. *The Great Derangement: Climate Change and the Unthinkable*. Chicago: University of Chicago Press.

Gibson, William. 1983. *Burning Chrome*. New York: Harper.

———. 1993. "Disneyland with the Death Penalty." *Wired*, April 1, 1993. https://www.wired.com/1993/04/gibson-2/.

———. 2012. *Distrust That Particular Flavor*. New York: Berkley Books.

Gidwani, Vinay. 2013. "Six Theses on Waste, Value, and Commons." *Social & Cultural Geography* 14 (7): 773–83.

Gilbert, Allen. 2007. "The Return of the Slum: Does Language Matter?" *International Journal of Urban and Regional Research* 31 (4): 697–713.

Goldman, Michael. 2011. "Speculative Urbanism and the Making of the Next World City." *International Journal of Urban and Regional Research* 34 (3): 551–81.

Goldstein, Jessica. 2018. *Planetary Improvement: Cleantech Entrepreneurship and the Contradictions of Green Capitalism*. Cambridge, MA: MIT Press.

Goodfellow, Tom. 2014. "Rwanda's Political Settlement and the Urban Transition: Expropriation, Construction and Taxation in Kigali." *Journal of Eastern African Studies* 8 (2): 311–29.

———. 2017. "Urban Fortunes and Skeleton Cityscapes: Real Estate and Late Urbanization in Kigali and Addis Ababa." *International Journal of Urban and Regional Research* 41:786–803.

———. 2022. *Politics and the Urban Frontier: Transformation and Divergence in Late Urbanizing East Africa*. New York: Oxford University Press.

Goodfellow, Tom, and Alyson Smith. 2013. "From Urban Catastrophe to 'Model' City? Politics, Security and Development in Post-Conflict Kigali." *Urban Studies* 50 (15): 3185–202.

Gordillo, Gastón. 2014. *Rubble: The Afterlife of Destruction*. Durham, NC: Duke University Press.

Gotham, Kevin, and Miriam Greenberg. 2014. *Crisis Cities: Disaster and Redevelopment in New York and New Orleans*. New York: Oxford University Press.

Grabski, Joanna. 2017. *Art World City: The Creative Economy of Artists and Urban Life in Dakar*. Bloomington: Indiana University Press.

Graham, Stephen, and Nigel Thrift. 2007. "Out of Order: Understanding Repair and Maintenance." *Theory, Culture & Society* 24 (3): 1–25.

Greenberg, Miriam. 2008. *Branding New York: How a City in Crisis Was Sold to the World*. New York: Routledge.

———. 2015. "'The Sustainability Edge': Competition, Crisis, and the Rise of Green Urban Branding." In *Sustainability in the Global City: Myth and Practice*, edited by Melissa Checker and Cynthia Isenhour, 105–30. Cambridge: Cambridge University Press.

Gregg, Emma. 2020. "Kigali: How Creativity Has Transformed the Rwandan Capital." *National Geographic*, March 8, 2020. https://www.nationalgeographic.com/travel/article/kigali-creativity-transformed-rwandan-capital.

Guardian. 2017. "Gods of Garbage—in Pictures." June 16, 2017. https://www.theguardian.com/artanddesign/gallery/2017/jun/16/gods-of-garbage-fabrice-monteiro-the-prophecy-polluted-environment-in-pictures.

Guma, Prince K. 2023. "Smart Cities and Their Settings in the Global South: Informality as a Marker." *Dialogues in Human Geography* 14 (3): 411–14.

Günel, Gökçe. 2019. *Spaceship in the Desert: Energy, Climate Change, and Urban Design in Abu Dhabi*. Durham, NC: Duke University Press.

Guyer, Jane. 2004. *Marginal Gains: Monetary Transactions in Atlantic Africa*. Chicago: University of Chicago Press.

Haggblade, Steven. 1990. "The Flip Side of Fashion: Used Clothing Exports to the Third World." *Journal of Development Studies* 26 (3): 505–21.

Hansen, Karen Tranberg. 1994. "Dealing with Used Clothing: Salaula and the Construction of Identity in Zambia's Third Republic." *Public Culture* 6:503–23.

———. 1999. "Secondhand Clothing Encounters in Zambia: Global Discourses, Western Commodities, and Local Histories." *Africa* 69 (3): 343–65.

———. 2000. *Salaula: The World of Secondhand Clothing and Zambia*. Chicago: University of Chicago Press.

Harms, Erik. 2016. *Luxury and Rubble: Civility and Dispossession in the New Saigon*. Oakland: University of California Press.

Harvey, David. 1989. "From Managerialism to Entrepreneurialism: The Transformation in Urban Governance in Late Capitalism." *Geografiska Annaler. Series B, Human Geography* 71 (1): 3–17.

———. 2003. *Paris: Capital of Modernity*. New York: Routledge.

Hecht, Gabrielle. 2018. "Interscalar Vehicles for an African Anthropocene: On Waste, Temporality, and Violence." *Cultural Anthropology* 33 (1): 109–44.

———. 2023. *Residual Governance: How South Africa Foretells Planetary Futures*. Durham, NC: Duke University Press.

Herbert, Claire, and Martin J. Murray. 2015. "Building from Scratch: New Cities, Privatized Urbanism, and the Spatial Restructuring of Johannesburg after Apartheid." *International Journal of Urban and Regional Research* 39 (3): 471–94.

Heyzer, Noeleen. 2014. "Mastering Our Urban Future." *Project Syndicate*, February 13, 2014. https://www.project-syndicate.org/commentary/noeleen-heyzer-on-the-six-steps-needed-to-create-livable-and-sustainable-cities.

Hickcox, Abby. 2007. "Greenbelt, White City: Race and Natural Landscape in Boulder, Colorado." *Discourse* 29:236–59.

———. 2017. "White Environmental Subjectivity and the Politics of Belonging." *Social & Cultural Geography* 19 (4): 496–519.

Hoffman, Danny. 2017. *Monrovia Modern: Urban Form and Political Imagination in Liberia*. Durham, NC: Duke University Press.

Holleman, Hannah. 2018. *Dustbowls of Empire: Imperialism, Environmental Politics, and the Injustice of "Green" Capitalism*. New Haven, CT: Yale University Press.

Holston, James. 1989. *The Modernist City: An Anthropological Critique of Brasilia*. Chicago: University of Chicago Press.

———. 2008. *Insurgent Citizenship: Disjunctions of Democracy and Modernity in Brazil*. Princeton, NJ: Princeton University Press.

Hopkinson, Nalo, and Uppinder Mehan. 2004. *So Long Been Dreaming: Postcolonial Science Fiction and Fantasy*. Sydney: Read How You Want.

Huchu, Tendai. 2012. "The Sale." In *AfroSF: Science Fiction by African Writers*, edited by Ivor W. Hartman, 32–41. Johannesburg: Storytime.

Hudani, Shakirah Esmail. 2020. "The Green Masterplan: Crisis, State Transition and Urban Transformation in Post-Genocide Rwanda." *International Journal of Urban and Regional Research* 44 (4): 673–90.

———. 2024. *Master Plans and Minor Acts: Repairing the City in Post-Genocide Rwanda*. Chicago: University of Chicago Press.

Hull, Matthew. 2012. *Government of Paper: The Materiality of Bureaucracy in Urban Pakistan.* Berkeley: University of California Press.

Human Rights Watch. 2003. *World Report 2003—Rwanda.* January 14, 2003. https://www.refworld.org/docid/3e28187b0.html.

———. 2006. *Swept Away: Street Children Illegally Detained in Kigali Rwanda.* May 2006. https://www.hrw.org/legacy/backgrounder/africa/rwanda0506/rwanda0506.pdf.

———. 2015. "'Why Not Call This Place a Prison?' Unlawful Detention and Ill-Treatment in Rwanda's Gikondo Transit Center." September 24, 2015. https://www.hrw.org/report/2015/09/24/why-not-call-place-prison/unlawful-detention-and-ill-treatment-rwandas-gikondo.

Iheka, Cajatan. 2018. *Naturalizing Africa: Ecological Violence, Agency, and Postcolonial Resistance in African Literature.* New York: Cambridge University Press.

———. 2021. *African Ecomedia: Network Forms, Planetary Politics.* Durham, NC: Duke University Press.

Ilberg, Antje. 2008. "Beyond Paper Policies: Planning Practice in Kigali." Paper presented at the 9th N-AERUS Conference, Heriot-Watt University, Edinburgh, Scotland, November 2008.

———. 2013. "Project Implementation, Constraints with Examples from Affordable Housing and Infrastructural Efforts." In *Cases on the Diffusion and Adoption of Sustainable Development Practices,* edited by Helen E. Muga and Ken D. Thomas, 147–66. Hershey, PA: IGI Global.

Ilberg, Antje, and Chris Rollins. 2007. *Low-Cost Construction Manual.* Engineers Without Borders USA.

Jackson, Steven J. 2014. "Rethinking Repair." In *Media Technologies: Essays on Communication, Materiality, and Society,* edited by Tarleton Gillespie, Pablo J. Boczkowski, and Kirsten A. Foot, 221–40. Cambridge, MA: MIT Press.

Jaganyi, Deogratius, Kato Njunwa, Manasse Nzayirambaho, P. Claver Rutayisire, Vincent Manirakiza, Aimable Nsabimana, Josephine Malonza, Mugabe Leon, and Gilbert Nduwayezu. 2018. *Rwanda: National Urban Policies and City Profiles for Kigali and Huye.* Glasgow: GCRF Centre for Sustainable, Healthy and Learning Cities and Neighbourhoods (SHLC), October 31, 2018. www.centreforsustainablecities.ac.uk/wp-content/uploads/2018/10/Research-Report-Rwanda-National-Urban-Policies-and-City-Profiles-for-Kigali-and-Huye.pdf.

Jefremovas, Villia. 2002. *Brickyards to Graveyards: From Production to Genocide in Rwanda.* Albany, New York: SUNY Press.

Jones, Branwen Gruffydd. 2009. "Cities without Slums? Global Architectures of Power and the African City." In *The African City Centre (Re)Sourced,* edited by Karel Bakker, 56–78. Pretoria: University of Pretoria.

Jones, Jeremy. 2010. "'Nothing Is Straight in Zimbabwe': The Rise of the *Kukiya-kiya* Economy 2000–2008." *Journal of Southern African Studies* 36 (2): 285–99.

Joyce, Patrick. 2003. *The Rule of Freedom: Liberalism and the Modern City.* New York: Verso.

Kagabo, José Hamim. 1982. *L'Islam et les "Swahili" au Rwanda.* Paris: École des Hautes Études en Sciences Sociales.

Kalisa, Armstrong Gatete. 2008. "Rwanda: Kiyovu Residents Decry Evictions." *Focus Media,* July 29, 2008. https://allafrica.com/stories/200807290996.html.

Kamuzinzi, Simon. 2016. "Barasaba gusubizwa ayo batanze ngo isoko rya Nyabugogo ryubakwe." *Kigali Today*, August 26, 2016. https://www.kigalitoday.com/amakuru/amakuru-mu-rwanda/barasaba-gusubizwa-ayo-batanze-ngo-isoko-rya-nyabugogo-ryubakwe.

Kandt, Richard. 1904. *Caput Nili: Eine empfindsame Reise zu den Quellen des Nils*. Berlin: D. Reimer.

Kanna, Ahmed. 2011. *Dubai: The City as Corporation*. Durham, NC: Duke University Press.

Kardish, Chris. 2014. "How Rwanda Became the World's Unlikely Leader in Plastic Bag Bans." *Governing: The Future of States and Localities*, April 18, 2014. https://www.governing.com/archive/gov-rwanda-plastic-bag-ban.html.

Kayiranga, Francis. 2018. "Aba bazahangana kugeza ryari? -> Kaboneka yabwiye abaturage ba 'Bannyahe' ati "abashaka guhangana bajye hariya duhangane." *Umuseke*, April 3, 2018. https://rugali.com/aba-bazahangana-kugeza-ryari-kaboneka-yabwiye-abaturage-ba-bannyahe-ati-abashaka-guhangana-bajye-hariya-duhangane.

Kedem, Shoshana. 2019. "Tony Blair's African Adventures Continue." *African Business*, November 11, 2019. https://african.business/2019/11/economy/tony-blairs-african-adventure-continues.

Kelley, Robin D. G. 2003. *Freedom Dreams: The Black Radical Imagination*. Boston: Beacon Press.

Kigali Institute of Science and Technology (KIST). 2001. *Kigali Economic Development Survey*.

Kigali Today. 2014. "Abacuruzi b'i Nyabugogo Bagiye Kuhubaka isoko rishya rya miliyari 32." *Kigali Today*, November 9, 2014. https://www.kigalitoday.com/ubukungu/ubucuruzi/article/abacuruzi-b-i-nyabugogo-bagiye-kuhubaka-isoko-rishya-rya-miliyari-32.

Kimari, Wangui. 2022. "'Colour ni Green': Ecological Futures in Nairobi Outlaw Style." *International Journal of Urban and Regional Research*, November 2022. Special online issue, "Spotlight on African Futures." https://www.ijurr.org/spotlight-on-african-futures/colour-ni-green-ecological-futures-in-nairobi-outlaw-style/.

Kimari, Wangui, and Henrik Ernstson. 2020. "Imperial Remains and Imperial Invitations: Centering Race within the Contemporary Large-Scale Infrastructures of East Africa." *Antipode* 52:825–46.

Kimenyi, Felly. 2008. "Rwanda: City Dwellers Receive Habitat Award in Style." *New Times*, October 17, 2008. https://allafrica.com/stories/200810200170.html.

Kimonyo, Jean-Paul. 2008. *Rwanda: Un Genocide Populaire*. Paris: Karthala.

Kinyanjui, Mary Njeri. 2013. "Women Informal Garment Traders in Taveta Road, Nairobi: From the Margins to the Center." *African Studies Review* 56 (3): 147–64.

———. 2014. *Women and the Informal Economy in Urban Africa*. New York: Zed Books.

Klein, Naomi. 1999. *No Logo: Taking Aim at the Brand Bullies*. New York: Picador.

———. 2007. *The Shock Doctrine: The Rise of Disaster Capitalism*. New York: Picador.

Kundabose, Jean David. 2013. "Imvura Idasanzwe Isize Rutikanga Ferdinand N'umuryango We Mu Kangaratete." *Inyarwanda*. September 12, 2013. https://inyarwanda.com/rw/amakuru/imvura-idasanzwe-isize-rutikanga-ferdinand-n-umuryango-we-mu-kangaratete-55160.

Kwizera, Prudence. 2022. "Muhanga: Abantu Batanu Bafatanywe Caguwa Ya Magendu." *Igihe*, September 30, 2022. https://mobile.igihe.com/amakuru/u-rwanda/article/muhanga-polisi-yafashe-abantu-batanu-binjije-mu-gihugu-imyenda-ya-caguwa.

Lamarque, Hugh. 2020. "Policing Small Communities: Rwandan Law Enforcement and the Co-production of Security." *Politique Africaine* 4 (160): 113–38.

Larkin, Brian. 2008. *Signal and Noise: Media, Infrastructure, and Urban Culture in Nigeria*. Durham, NC: Duke University Press.

Lemarchand, René. 1970. *Rwanda and Burundi*. New York: Praeger Publishers.

———. 2010. "The Apocalypse in Rwanda." *Cultural Survival*, March 19, 2010. https://www.culturalsurvival.org/publications/cultural-survival-quarterly/apocalypse-Rwanda.

Longman, Timothy. 2017. *Memory and Justice in Post-genocide Rwanda*. New York: Cambridge University Press.

Louis, William Roger. 1963. *Ruanda-Urundi: 1884–1919*. Oxford: Clarendon Press.

Luvaas, Brent. 2012. *DIY Style: Fashion, Music, and Global Digital Cultures*. New York: Berg.

Macaskill, Andrew. 2023. "UK Estimates Cost of Deporting Each Asylum Seeker to Rwanda Will Be 169,000 Pounds." *Reuters*, June 26, 2023. https://www.reuters.com/world/uk/uk-says-cost-deporting-each-asylum-seeker-rwanda-be-169000-pounds-2023-06-26/.

Makhulu, Anne-Maria. 2015. *Making Freedom: Apartheid, Squatter Politics, and the Struggle for Home*. Durham, NC: Duke University Press.

Mamdani, Mahmood. 2001. *When Victims Become Killers: Colonialism, Nativism, and the Genocide in Rwanda*. Princeton, NJ: Princeton University Press.

Manirakiza, Vincent. 2011. "Processus d'urbanisation de la Ville de Kigali, Rwanda: Relation entre la dynamique spatiale et démographique." Communication pour la Chaire Quetelet Urbanisation, Migrations Internes et Comportements Démographiques.

Manirakiza, Vincent, and An Ansoms. 2014. "Modernizing Kigali: The Struggle for Space in the Rwandan Urban Context." In *Losing Your Land: Dispossession in the Great Lakes Region*, edited by An Ansoms and Thea Hilhorst, 186–203. Rochester, NY: James Currey.

Manirakiza, Vincent, Leone Mugabe, Aimable Nsabimana, and Manassé Nzayirambaho. 2019. "City Profile: Kigali, Rwanda." *Environment and Urbanization ASIA* 10 (2): 290–307.

Marshall, Lisa. 2008. "Rebuilding Rwanda." *ColoradoBiz*, April 2008. www.lisaannmarshall.com/Stories/Rwanda.pdf.

Mbembe, Achille, and Sarah Nuttall, eds. 2008. *Johannesburg: The Elusive Metropolis*. Durham, NC: Duke University Press.

McKittrick, Katherine. 2006. *Demonic Grounds: Black Women and the Cartographies of Struggle*. Minneapolis: University of Minnesota Press.

Meintjes, Louise. 2017. "Hi-Fi Sociality, Lo-Fi Sound: Affect and Precarity in an Independent South African Recording Studio." In *State and Culture in Postcolonial Africa: Enchantings*, edited by Tejumola Olaniyan, 207–23. Bloomington, IN: Indiana University Press.

Melly, Caroline. 2017. *Bottleneck: Moving, Building, and Belonging in an African City*. Chicago: University of Chicago Press.

Merrifield, Andy. 2000. "The Dialectics of Dystopia: Disorder and Zero-Tolerance in the City." *International Journal of Urban and Regional Research* 24 (2): 473–89.

Michalski, Wenzel. 2014. "A Quiet Murder in Rwanda." *Human Rights Watch*, July 20, 2014. https://www.hrw.org/news/2014/07/20/quiet-murder-rwanda.

MINECOFIN. 2023. "Kigali's Informal Settlements to Be Upgraded through AFD Support." Accessed January 1, 2024. https://www.minecofin.gov.rw/news-detail/kigalis-informal-settlements-to-be-upgraded-through-afd-support.

Ministry of Infrastructure. 2015. National Housing Policy. Government of Rwanda, Kigali.

Mitchell, Karen. 2005. "Into Africa: Boulder Architects to Design Master Plan for Kigali, Rwanda." *Daily Camera*, June 6, 2005, 1–5.

Mitchell, Timothy. 2002. *Rule of Experts: Egypt, Techno-Politics, Modernity*. Berkeley: University of California Press.

Moonsamy, Nedine. 2016. "Life Is a Biological Risk: Contagion, Contamination, and Utopia in African Science Fiction." *Cambridge Journal of Postcolonial Literary Inquiry* 3 (3): 329–43.

Moore, Jason. 2015. *Capitalism in the Web of Life*. New York: Verso.

Moreno, Eduardo López. 2003. *Slums of the World*. New York: UN-Habitat.

Morscher, Livia, and Timo Nothdurft. 2021. "1000 Hills, 1 Plan." *Forbes*, February 2, 2021. https://www.forbes.at/artikel/1000-hills-1-plan.html.

Morton, David. 2019. *Age of Concrete: Housing and the Shape of Aspiration in the Capital of Mozambique*. Athens, OH: Ohio University Press.

Moyo, Dambisa. 2009. *Dead Aid: Why Aid Is Not Working and Why There Is a Better Way Forward*. New York: Farrar, Straus and Giroux.

Mugisha, John. 2015. "Compensation for Land Expropriation in Rwanda: The Need for Conventional Approaches to Valuation." *Journal of Land Administration in Eastern Africa* 3:296–306.

Murray, Martin. 2015. "City Doubles: Re-Urbanism in Africa." In *Cities and Inequalities in a Global and Neoliberal World*, edited by Faranak Miraftab, David Wilson, and Ken Salo, 92–109. New York: Routledge.

Mususa, Patience. 2021. *There Used to Be Order: Life on the Copperbelt after the Privatisation of the Zambia Consolidated Copper Mines*. Ann Arbor: University of Michigan Press.

Mutongi, Kenda. 2017. *Matatu: A History of Popular Transportation in Nairobi*. Chicago: University of Chicago Press.

Mwai, Collins. 2015. "Kigali City Brand Image Vital: Ex-Seoul Mayor." *New Times*, January 4, 2015. https://www.newtimes.co.rw/article/114749/News/kigali-city-brand-image-vital-ex-seoul-mayor.

Myers, Garth. 2005. *Disposable Cities: Garbage, Governance and Sustainable Development in Urban Africa*. Burlington, VT: Ashgate Publishing.

———. 2011. *African Cities: Alternative Visions of Theory and Practice*. New York: Zed Books.

———. 2016. *Urban Environments in Africa: A Critical Analysis of Environmental Politics*. Bristol, UK: Policy Press.

Nakassis, Constantine V. 2013. "Brands and their Surfeits." *Cultural Anthropology* 28:111–26.

Neuwirth, Robert. 2021. "A Singaporean Firm Has Become the Go-To Master Planner for African Cities." *City Monitor*, August 20, 2021. https://city2city.network/singaporean-firm-has-become-go-master-planner-african-cities.

Newbury, Catharine. 1988. *The Cohesion of Oppression: Clientship and Ethnicity in Rwanda, 1860–1960*. New York: Columbia University Press.

Newell, Stephanie. 2020. *Histories of Dirt: Media and Urban Life in Colonial and Postcolonial Lagos*. Durham, NC: Duke University Press.

Ngabonziza, Justin. 2023. "Kigali: Abazunguzayi Bishe Umunyerondo Bamuteye Icyuma." *Inyarwanda*, March 8, 2023. https://inyarwanda.com/inkuru/126823/kigali-abazunguzayi-bishe-umunyerondo-bamuteye-icyuma-126823.html.

Nikuze, Alice, Richard Sliuzas, and Johannes Flacke. 2020. "From Closed to Claimed Spaces for Participation: Contestation in Urban Redevelopment Induced-Displacements and Resettlement in Kigali, Rwanda." *Land* 9 (7): 212.

Nikuze, Alice, Richard Sliuzas, Johannes Flacke, and Martin Van Maarseveen. 2019. "Livelihood Impacts of Displacement on Informal Households—A Case Study from Kigali, Rwanda." *Habitat International* 86:38–47.

Nixon, Rob. 2011. *Slow Violence and the Environmentalism of the Poor*. Cambridge, MA: Harvard University Press.

Niyomwungeri, Cyprien. 2017. "Abazunguzayi n'utujagari mu byashegeshe Umujyi wa Kigali Mu Myaka Irindwi Ishize." *Igihe*, July 6, 2017. https://igihe.com/amakuru/u-rwanda/article/abazunguzayi-n-utujagari-mu-byashegeshe-umujyi-wa-kigali-mu-myaka-irindwi.

Niyoyita, Joseline, and Haiying Li. 2019. "Study on Rwandan Traditional Architecture." *Advances in Social Science Education and Humanities Research* 324:185–89.

Nsabimana, Natacha. 2023. "Genocide-time: Political Violence Reckoning in Rwanda." *American Anthropologist* 125:761–70.

Ntakiyimana, Charles, Yu Zhang, and Gratien Twagirayezu. 2022. "Road Flooding in Kigali City, Rwanda: Causes, Effects on Road Transportation and Mitigation Measures." *Polish Journal of Environmental Studies* 31 (4): 3735–44.

Okorafor, Nnedi. 2019. "Africanfuturism Defined." https://nnedi.blogspot.com/2019/10/africanfuturism-defined.html.

Olson, Jennifer Maria. 1990. "The Impact of Socioeconomic Factors on Migration in Rwanda." MA thesis, Michigan State University.

Omelsky, Matthew. 2014. "After the End Times: Post-Crises African Science Fiction." *Cambridge Journal of Postcolonial Literary Inquiry* 1 (1): 33–49.

OZ Architecture. 2007. The Kigali Conceptual Master Plan.

Patel, Raj, and Jason Moore. 2018. *A History of the World in Seven Cheap Things: A Guide to Capitalism, Nature, and the Future of the Planet*. Berkeley: University of California Press.

Pettem, Sylvia. 2010. "Boulder Population Nearly Doubled in the 1950s." *Boulder Daily Camera*, January 12, 2010. https://www.dailycamera.com/2010/01/12/boulder-population-nearly-doubled-in-the-1950s/.

Pierre, Jemima. 2012. *The Predicament of Blackness: Postcolonial Ghana and the Politics of Race*. Chicago: University of Chicago Press.

Piot, Charles. 1999. *Remotely Global: Village Modernity in West Africa*. Chicago: University of Chicago Press.

———. 2010. *Nostalgia for the Future: West Africa after the Cold War*. Chicago: University of Chicago Press.

Piot, Charles, and Kodjo Nicolas Batema. 2019. *The Fixer: Visa Lottery Chronicles*. Durham, NC: Duke University Press.

Planet Consortium. 2012. "Housing Market Demand, Housing Finance, and Housing Preferences for the City of Kigali."

Planet Money. 2013. "Planet Money's T-Shirt Project." NPR, December. https://www.npr.org/series/248799434/planet-moneys-t-shirt-project.

Powanga, Luka, and Paul Adjei Kwakwa. 2024. "Determinants of Carbon Emissions in Kenya and Policy Implications." *Journal of Environmental Management* (370). https://doi.org/10.1016/j.jenvman.2024.122595.

Purdeková, Andrea. 2011. "Even When I Am Not Here, There Are So Many Eyes: Surveillance and State Reach in Rwanda." *Journal of African Studies* (49) 3: 475–97.

———. 2015. *Making Ubumwe: Power, State, and Camps in Rwanda's Unity-Building Project*. New York: Berghahn Books.

Purdeková, Andrea, and David Mwambari. 2021. "Post-Genocide Identity Politics and Colonial Durabilities in Rwanda." *Critical African Studies* 14 (1): 19–37.

Quayson, Ato. 2014. *Oxford Street, Accra: City Life and the Itineraries of Transnationalism*. Durham, NC: Duke University Press.

Rabinow, Paul. 1989. *French Modern: Norms and Forms of the Social Environment*. Chicago: University of Chicago Press.

Reuters. 2025. "US Deports Iraqi Man at the Centre of Debate on Refugee Policy." April 24, 2025. https://www.reuters.com/world/us-deports-iraqi-man-centre-debate-refugee-policy-2025-04-24/.

Rieder, John. 2008. *Colonialism and the Emergence of Science Fiction*. Middletown, CT: Wesleyan University Press.

Rivkin Associates Inc. 1983. "Report on a Field Investigation on Settlement Patterns and Housing in the Republic of Rwanda." December 1983.

Robinson, Jennifer. 2002. "Global and World Cities: A View from Off the Map." *International Journal of Urban and Regional Research* 26:531–54.

———. 2006. *Ordinary Cities: Between Modernity and Development*. New York: Routledge.

Rodgers, Lucy. 2018. "Climate Change: The Massive CO2 Emitter You May Not Know About." *BBC Online*, December 17, 2018. https://www.bbc.com/news/science-environment-46455844.

Roitman, Janet. 2014. *Anti-Crisis*. Durham, NC: Duke University Press.

Roy, Ananya. 2010a. *Poverty Capital: Microfinance and the Making of Development*. New York: Routledge.

———. 2010b. "Informality and the Politics of Planning." In *The Ashgate Research Companion to Planning Theory: Conceptual Challenges for Spatial Planning*, edited by Patsy Healey and Jean Hillier, 87–107. New York: Routledge.

Rubinoff, Donna. 2014. "Rwanda Leading the Way to Sustainable Urbanism." *New Times*, March 18, 2014. https://www.newtimes.co.rw/article/104956/Opinions/rwanda-leading-the-way-to-sustainable-urbanism/amp.

Rutheiser, Charles. 1996. *Imagineering Atlanta: The Politics of Place in the City of Dreams*. New York: Verso.

Ruxin, Joshua. 2013. *A Thousand Hills to Heaven: Love, Hope, and a Restaurant in Rwanda*. New York: Little Brown and Company.

Rwanda Environment Management Authority. 2013. "Kigali: State of the Environment Outlook Report."

Rwanda National Police. 2023. "Smugglers Intercepted with Bales of Used Clothes." RNP, July 27, 2023. https://police.gov.rw/media/news-detail/news/smugglers-intercepted-with-bales-of-used-clothes/.

Rwanda Updates. 2022. "Amarira N'imiborogo Abaturage Ba Kangondo Bari Kwimurwa Kugahira Leta Yohereje Batayo Zabapolis." September 15, 2022. News video, 20:12. https://www.youtube.com/watch?v=4huA00WMCCk.

Rwirahira, Rodrigue. 2012. "Rwanda BRD Finances Modern Market in Gisozi." *Rwanda Focus*, January 11, 2012. https://allafrica.com/stories/201201131104.html.

Schramm, Sophie, and Basil Ibrahim. 2021. "Hacking the Pipes: Hydro-political Currents in a Nairobi Housing Estate." *Environment and Planning C: Politics and Space* 39(2): 354–70. https://doi.org/10.1177/2399654419865760.

Scott, James. 1998. *Seeing like a State: How Certain Schemes to Improve the Human Condition Have Failed.* New Haven, CT: Yale University Press.

Shabazz, Rashad. 2015. *Spatializing Blackness: Architectures of Confinement and Black Masculinity in Chicago.* Champaign: University of Illinois Press.

Shearer, Samuel. 2015. "Producing Sustainable Futures in Post-Genocide Kigali, Rwanda." In *Sustainability in the Global City: Myth and Practice*, edited by Melissa Checker and Cynthia Isenhauer, 180–86. Cambridge: Cambridge University Press.

———. 2020a. "The City Is Burning! Street Economies and the Juxtacity of Kigali, Rwanda." *Urban Forum* 31:351–71.

———. 2020b. "Revanchist Kigali: Retro-Victorian Urbanism and the Gentrification of a Twenty-First-Century Metropolis." In *Gentrification around the World*, vol. 2, edited by Jerome Krase and Julie DeSena, 189–217. New York: Palgrave.

———. 2024. "'This Place Is Fake': Green Capitalism and the Production of Scarcity in Kigali, Rwanda." *City & Society* 36 (3): 146–59.

Simone, AbdouMaliq. 2004a. *For the City Yet to Come: Changing Life in Four African Cities.* Durham, NC: Duke University Press.

———. 2004b. "People as Infrastructure: Intersecting Fragments in Johannesburg." *Public Culture* 16 (3): 407–29.

———. 2009. *City Life from Jakarta to Dakar: Movements at the Crossroads.* New York: Routledge.

———. 2010. "A Town on Its Knees?" *Theory, Culture & Society* 27 (7–8): 130–54.

Singirankabo, Edouard, and Emmanuel Iyamuremye. 2022. "Modelling Extreme Rainfall Events in Kigali City Using Generalized Pareto Distribution." *Meteorological Applications* 29 (4): e2076. https://rmets.onlinelibrary.wiley.com/doi/full/10.1002/met.2076.

Sirven, Pierre. 1984. *La sous urbanisation et les villes du Rwanda et du Burundi.* Thèse de Doctorate en Géographie, University of Bordeaux III.

Smith, Korydon H., Tomà Berlanda, Stephen Luoni, and the University of Arkansas Community Design Center. 2018. *Interpreting Kigali, Rwanda: Architectural Inquiries and Prospects for a Developing African City.* Fayetteville: University of Arkansas Press.

Smith, Neil. (1996) 2005. *The New Frontier: Gentrification and the Revanchist City.* London: Routledge.

Smith, P. 2023. "Value of the Secondhand Apparel Market Worldwide from 2021 to 2027." *Statista*, September 5. https://www.statista.com/statistics/826162/apparel-resale-market-value-worldwide/.

Sommers, Marc. 2012. *Stuck: Rwandan Youth and the Struggle for Adulthood.* Athens: University of Georgia Press.

Soy, Anna. 2022. "UK-Rwanda Asylum Seekers' Deal: Good News for Kigali Hotels." *BBC*, May 20, 2022. https://www.bbc.com/news/world-africa-61496397.

Star, Susan Leigh. 1999. "The Ethnography of Infrastructure." *American Behavioral Scientist* 43 (3): 377–91.

Stoner, Jill. 2012. *Toward a Minor Architecture.* Cambridge, MA: MIT Press.

Strauss, Scott, and Lars Waldorf, eds. 2011. "Introduction: Seeing Like a Post-Conflict State." In *Remaking Rwanda: State Building and Human Rights after Mass Violence.* Madison: University of Wisconsin Press.

Striebig, Bradley, Adebayo A. Ogundipe, and Maria Papadakis. 2015. *Engineering Applications in Sustainable Design and Development*. Boston: Cengage Learning.

Sugrue, Thomas. 1996. *The Origins of the Urban Crisis: Race and Inequality in Postwar Detroit*. Princeton, NJ: Princeton University Press.

Surbana, LLC. 2010. *Final Nyuragenge Report*.

Surbana, LLC. 2012. "The City of Kigali: Towards a Centre of Urban Excellence." Presentation to the City of Kigali, March 2, 2012.

Surbana-Jurong and City of Kigali. 2013. The Kigali Master Plan.

Suvin, Darko. 1979. *Metamorphosis of Science Fiction: On the Politics and History of a Literary Genre*. New Haven, CT: Yale University Press.

Swyngedouw, Eric. 2009. "The Antinomies of the Postpolitical City: In Search of a Democratic Politics of Environmental Production." *International Journal of Urban and Regional Research* 33 (3): 601–20.

Taylor, Keeanga-Yamahtta. 2019. *Race for Profit: How Banks and the Real Estate Industry Undermined Black Home Ownership*. Chapel Hill: University of North Carolina Press.

Topping, Alexandra. 2014. "Kigali's Future or Costly Fantasy? Plan to Reshape Rwandan City Divides Opinion." *Guardian*, April 4, 2014. https://www.theguardian.com/world/2014/apr/04/kigali-plan-rwandan-city-afford-new-homes-offices.

Trouillot, Michel-Rolph. 1995. *Silencing the Past: Power and the Production of History*. New York: Beacon Press.

Tsamaase, Tlotlo. 2020. "Behind Our Irises." In *Africanfuturism: An Anthology*, edited by Wole Talabi, 42–51. *Brittle Paper*. chrome-extension://efaidnbmnnnibpcajpcglcle findmkaj/https://brittlepaper.com/wp-content/uploads/2020/10/Africanfuturism-An-Anthology-edited-by-Wole-Talabi.pdf.

Tsing, Anna Lowenhaupt. 2012. "On Non-Scalability: The Living World Is Not Amenable to Precision Nested Scales." *Common Knowledge* 18 (3): 505–24.

———. 2015. *Mushroom at the End of the World: On the Possibility of Life in Capitalist Ruins*. Princeton, NJ: Princeton University Press.

Tuyishimire, Raymond. 2022. "Shikama Winangiye Kwimuka 'Bannyahe' Yasabiwe gufungwa iminsi 30 Y'agateganyo." *Umuseke*, September 22, 2022. https://umuseke.rw/2022/09/shikama-winangiye-kwimuka-bannyahe-yasabiwe-gufungwa-iminsi-30-yagateganyo/.

Twahirwa, Aimable. 2018. "Cleanest City in Africa? Kigali Scrubs Up." Reuters, April 21, 2018. https://www.reuters.com/article/world/cleanest-city-in-africa-kigali-scrubs-up-idUSK BN1HR2F8.

Twizeyimana, Fabrice. 2014. "Gasabo: Bagujije Akayabo K'amafaranga Bubaka Inzu Z'ubucuruzi, None Abakirya ni Mbarwa." *Igihe*, June 8, 2014. https://www.igihe.com/ubukungu/ubucuruzi/article/gasabo-bagujije-akayabo-k.

Ukwezi TV. 2016. "Imyogaragambyo Y'abazunguzayi Bamagana Mugenzi Wabo Whaohotewe." June 10, 2016. News video, 1:34. https://www.youtube.com/watch?v=JRlovk D2-pk.

———. 2022. "Kangondo: Tuatera Akabariro Gute Turyamanye N'abana? Mwibeshyera Perezida Kagame Si We Uturenganya." September 8, 2022. News video, 32:59. https://www.youtube.com/watch?v=doe7jxmYueo.

Umuseke. 2014. "Nyabugogo: Abacuruzi Ntibumva Impamvu Bimuwe Huti Huti." *Umuseke*, October 16, 2014.

Umutoni, Louise. 2016. "Establishing a New Literary Award Prize: The Huza Press Award for Fiction." *Africa in Words Guest*, November 27, 2016. https://africainwords.com/2016/11/27/establishing-a-new-literary-prize-the-huza-press-award-for-fiction/.

UN Department of Economic and Social Affairs. 2023. "11: Make Cities and Human Settlements Inclusive, Safe, Resilient and Sustainable." Sustainable Development Goals, https://sdgs.un.org/goals/goal11.

UN-Habitat. 2003. *The Challenge of Slums*. https://unhabitat.org/sites/default/files/download-manager-files/The%20Challenge%20of%20Slums%20-%20Global%20Report%20on%20Human%20Settlements%202003.pdf.

Urban Boyz and Jay Polly. 2014. "Caguwa." Music video, 3:52. Amakuru. https://www.youtube.com/watch?v=10cXUZ853Jc.

US Census Bureau. n.d. QuickFacts, Boulder County, Colorado. Table RHI125221 (December 27, 2023). https://www.census.gov/quickfacts/fact/table/bouldercountycolorado/RHI125221.

Uvin, Peter. 1998. *Aiding Violence: The Development Enterprise in Rwanda*. West Hartford, CT: Kumarian Press.

Uwayezu, Ernest, and Walter T. de Vries. 2020a. "Access to Affordable Houses for the Low-Income Urban Dwellers in Kigali: Analysis Based on Sale Prices." *Land* 9 (3): 85.

———. 2020b. "Can In-Kind Compensation for Expropriated Real Property Promote Spatial Justice? A Case Study Analysis of Resettlement in Kigali City, Rwanda." *Sustainability* 12 (9): 3753.

Uzeyman, Anisia, and Saul Williams, dirs. 2021. *Neptune Frost*. Film. New York: Knitting Factory Entertainment.

Vandersypen, Marijke. 1977. "Femmes Libres de Kigali." *Cahiers d'études Africaines* 17 (65): 95–120.

Vansina, Jan. 2004. *Antecedents to Modern Rwanda: The Nyiginya Kingdom*. Madison: University of Wisconsin Press.

Vergès, Françoise. 2017. "Racial Capitalocene." In *Futures of Black Radicalism*, edited by Gaye Theresa Johnson and Alex Lubin, 72–82. London: Verso.

Von Schnitzler, Anita. 2016. *Democracy's Infrastructure: Technopolitics and Protest after Apartheid*. Princeton, NJ: Princeton University Press.

Vumbi, Danny. 2014. "Ni Danger." Music video, 4:35. Ireview Live. https://www.youtube.com/watch?v=xFi3szI8_7s.

Wakhungu, Judy, Chris Huggins, Elvin Nyukuri, and Jane Lumumba. 2010. "Approaches to Urban Informal Settlements in Africa: Experiences from Kigali and Nairobi, a Policy Brief." *Africa Center for Technology Studies*.

Wallace, Paul, and Lyubov Pronina. 2014. "Rwanda May Sell Next Dollar Bond with Growth Beating Region." *Bloomberg*, October 20, 2014. https://www.bloomberg.com/news/articles/2014-10-20/rwanda-may-sell-dollar-bond-next-year-with-growth-beating-region.

Watson, Vanessa. 2014. "African Urban Fantasies: Dreams or Nightmares?" *Environment and Urbanization* 26 (1): 215–31.

Williams, Florence. 2008. "Twenty-Five Square Miles Surrounded by Reality." *New York Times*, March 30, 2008. https://www.nytimes.com/2008/03/30/style/tmagazine/30boulder.html.

Woods, Maxwell. 2019. "Punk Urbanism: Insurgency, Crisis, and Cultural Geography." *Social & Cultural Geography* 22 (5): 666–85.

World Bank. 1972. "Announcement of Three Million Dollars Credit for Highway Maintenance." IDA Press Release 72/73. March 23, 1972. https://documents.worldbank.org/en/publication/documents-reports/documentdetail/566471600946199901/announcement-of-three-million-dollars-credit-to-rwanda-for-highway-maintenance-on-march-23-1972.

———. 1979. *Project Performance Audit Report: Rwanda's First, Second and Third Highway Projects*. Operations Department, June 13, 1979. https://documents1.worldbank.org/curated/en/185001468915317272/pdf/multipage.pdf.

World Bank Group. 2020. *Economy Profile: Rwanda*. Doing Business 2020. https://www.doingbusiness.org/content/dam/doingBusiness/country/r/rwanda/RWA.pdf.

Wright, Gwendolyn. 1991. *The Politics of Design in French Colonial Urbanism*. Chicago: University of Chicago Press.

GENERAL INDEX

abadasso, 121, 131, 136
abadozi, 142–43, 144*fig.*
abafundi, 26, 41, 46, 83, 92, 111, 114–15, 130, 137, 179, 186n1
abanyerondo, 115, 121, 123, 130, 132, 136
abapandagari, 136
abashoramari, 142
abaturage, xi, 27, 103. See also *umuturage*
abayedi, 130
abayobozi, 42, 111, 114–15. See also *umuyobozi*
abazunguzayi, 113, 115, 126–33, 139, 141–42, 148, 177. See also *umuzunguzayi*
Abu Dhabi, 61
Accra, 8, 12, 127
activism: artistic ecological, 153–54; environmental, 90, 188n3; history of unionism or, 25. See also environmentalists; environmental justice
Addis Ababa, 8
aesthetics, 152; "aesthetic dissent," 103; market-driven sustainable green, 165; minimalist, 174; trash, 153–58, 165–66, 179
African cities: discourses relating to, 4, 56, 91; perceptions of deficiency in, 90–91, 179; proponents of green growth relating to, 11. See also urbanism
African cyberpunk, 24, 116–18. See also Africanfuturism
Africanfuturism, 4, 9, 16, 21–22, 24, 104, 155, 174–79; as expressive culture in postcolonial literature, 21; as punk, 132. See also African cyberpunk
African urban studies, 116, 166
Afro-Arabs, 35
agatare level, 83
agriculture: linear, 76; modernization programs for, 44; three-pond filtration system near the wetlands to convert effluent into fertilizer for, 114, 138
akajagari, 5, 18, 22, 27, 95–96, 111–12, 115–24, 131–37, 140–41, 157, 171, 189n4
akantu/akafanta, 51, 136
akarenganye, 18, 108
amadeni, 149
amaduka, 148
AmaG the Black: "Nyabarongo" (music video), xi
amatafari yahiye, 39
amatafari ya rukarakara, 38, 38*fig.*, 41, 49, 98, 99*fig.*, 102, 105–6, 112, 119, 124, 171, 179, 188n7
amatongo gusa, 109, 146
Anand, Nikhil, 173
Ansoms, An, 137
apartheid, 17, 45, 117. See also South Africa
Appel, Hannah, 91
ArcGIS databases, 20, 80
architectures: colonial administrators and missionaries as fascinated by Rwandan, 35; colonial durabilities and hybrid, 34–39; of concrete, steel, and glass, 98, 104, 109; histories of clay, 24, 37; histories of colonial and postcolonial household, 37; imperfect,

architectures *(continued)*
24, 95–101, 111, 138, 174, 179; independence-era residents using colonial-era hybrid clay, 28; processes of land transfer and, 42, 50, 53; twentieth-century modernist, 61–62, 66. *See also* built environments; housing; urban planning
Askari, 32–33
Astrida (Huye), 33

bakavukire, 111, 116
Bannyahe, xiv, 56, 96, 112–41, 188n1, 189n4; demolition in, 122, 138–41, 146; electricity poles installed in the lower section of, 125*fig.*; government services for, 114–15; repair of a neighborhood that was supposed to remain broken in, 118, 123; residents capitalizing on Kigali's infrastructure in connecting, 123–26; satellite images of, 120*fig.*; "sustainable" luxury housing estate (Savannah Creek) to be built over, 138–41, 170; waste management system of, 114. *See also* neighborhoods
Batsinda estate, 18–19, 86–91, 93, 96, 100–112, 119, 139, 169, 171; brightly colored bungalows built on the location of the, 110*fig.*; as "failure" of green capitalism, 112; as retrofitted with electricity, 110. *See also* housing
Bedore, Pamela, 15–16
Belgians, 32–34, 46
Belgium, 147
Beukes, Lauren: *Zoo City*, 22, 117–18, 176–78
biodiversity, 20
Bissell, William, 30, 32
Blair, Tony, 56, 84, 188n7
boma, 30, 32, 35
Botswana, 104
Boulder, xii, xiii, 64–65, 68, 76–77
brand image, 12–14, 24, 55–85; catastrophes and, 64–68; as postcrisis city, 7, 71–73, 79, 89. *See also* urban planning
Brasilia, 61
bribes, 82–83, 111; strategic, 135–37. *See also* corruption
B-Threy: "Nituebue/Nitwebwe" (music video), 164–65
building codes, 40, 44–45; and informality, 98; stringent new, 82; as unenforceable, 173; widespread violations of, 88, 134; zoning and, 8, 83, 100, 121, 134. *See also* zoning
built environments: accounting for existing conditions of the, 72; as assembled from materials that have been broken down and repurposed, 117–20; commitment to minimalism and functionality over aesthetics in the working-class, 97–99; history of segregated, 53; landowners and lower-level state officials as engaged with their, 46; for producing colonial subjects, 37; production of scarcity and the destruction of, 3–4, 88–92; satellite base map of, 81. *See also* architectures; construction; repair; urban planning
Bukoba, 32
Burundi, 33; Kigali residents in refugee camps in, 48–49; as placed under Belgian "tutelage" by a League of Nations mandate, 32
Burundians, 4, 6, 40
Bushali, 174; "Kivuruga" (music video), 165; "Nituebue/Nitwebwe" (music video), 164–65; "Yarababye" (music video), 165
Butler, Octavia, 174

caguwa, 143–57, 143*fig.*, 161–62, 164, 179; economic precarity as part of, 166; smuggling of, 190n3; taxed and reported imports of, 190n3. *See also* fashion
Cairo, 12
Canada, 65
Cape Town, 10, 17, 45. *See also* South Africa
capitalism: alienating market logics of corporate, 116; catastrophe and, 67, 169; Cold War, 47, 63; crises generated by, 21, 175; market growth and zero-carbon, 10, 63; postcolonial studies and racial, 77; wasting processes of, 158. *See also* climate change; green capitalism; neoliberalism; urbanism
Capitalocene, 19, 169
Checker, Melissa, 11, 191n2
Chicago School economists, 67
China, 66, 147
Choice Live, The (Kinyarwanda vlog), 164
cholera, 41
Cities Alliance, 91
clay, 35, 37–41, 71, 82, 92; as cheap building material, 104–5, 114, 119; contraband bricks made from, 111; disallowing houses built with bricks of, 100; as earthquake and fire resistant, 98; harvesting of, 186n1; impermanence of, 99; Kigali as city of, 39–41, 46, 50, 99; malleability of, 102. *See also* construction
climate change: capital-induced, 21; and "durable" materials like concrete, 98; Kigali as green city that can withstand the shocks

of, 3, 58; as salient international issue, 10; as site of production, 174; and "technical adjustments," 11. *See also* capitalism; environmentalists; greenhouse gas emissions

Cocomanya, 160–62

coffee industry, 44, 47

Cold War, 39, 47

colonialism, xv, 7, 28–29; cheap, 32; efforts to contain Kigali during, 52; foundation of division and conflict in Rwanda from, 34; harms of capitalism and the residues of, 41; history of settler, 65

Congo, 29, 32, 40, 46–49; refugee camps for Rwandans in the, 48–49. *See also* Kinshasha

Congolese, 4, 6, 40, 149

conservation, 64–65; carbon-free tourism as, 151; urban maintenance and, 174; wetland, 70, 74. *See also* ecologies; environmentalists; waste

construction: local, 101–2; low-impact, low-carbon technologies in, 89, 100; projects of, 39–41, 91, 112; unauthorized, 121, 123, 137. *See also* built environments; clay; master plan projects; repair

consultants, 40, 60, 67, 72, 74, 81, 89–92, 173; United Kingdom, 80. *See also* design firms; Kigali City Master Plan; OZ Architecture; Surbana

consumption, 155; Eurocentric media models of high, 97; Global North centers of, 146; high-stakes labor of transforming clothing into fashion through the work of, 153; and social reproduction, 103

corruption, 83–84, 136, 188n6; state violence against the street economy and, 131. *See also* bribes

Cyangugu, 40, 42

Dakar, 8, 17, 158; Mbeubeuss waste dump in, 156

Dar es Salaam, 8, 17, 30, 32, 125. *See also* Tanzania

de Boeck, Filip, 45, 152

decentralization, 51; of surveillance, 27

decolonization, 45–46, 186n4; postindependence urban, 28, 46; reforms promised by, 45

Delhi, 61

democracy, 8; Kigali envisioned as a new model of urban, 10

demolitions: in Bannyahe, 122, 138–41, 146; by Belgian colonial authorities, 33; costly spectacle of neighborhood, 6, 18–19, 136; discourses that justify, 21, 23, 74, 83, 112, 139, 167; housing scarcity as generated through, 100–101, 112; in Kimicanga, 1, 4, 6–7, 12, 17, 56, 85, 110, 120, 128, 159, 167, 169, 185n6; in Kiyovu cy'abakene, xiii, 58, 78–88, 92, 96, 101, 106–12, 119–20, 133, 137, 159, 169, 171–72; master plan, xiii, 78–79, 122–23, 133, 138, 140, 167, 169, 189n4; of Nyabugogo Market, 154, 158; for rebuilding Kigali as privatized green metropolis, 5, 12, 56, 74, 78–79, 81, 85, 89, 100, 118–21, 138–39. *See also* evictees; housing; informal settlements; master plan projects

Denver Tech Center, 10, 60–61, 66

design firms, 3, 91; global, 7–8, 23. *See also* consultants; urban planning

détournement, 121, 189n7

Dickson, Jessica, 118, 176–77

digital technology and green industries, 166

displacement: compensation costs of dispossession and, 23, 110; discourses that justify, 21; injustices of, 18, 86–88, 106; master plan demolitions and, 5, 12, 56, 74, 78–79, 81, 110, 172; and visions of urban development and futurity, 146. *See also* demolitions; reparations

Doherty, Jacob, 40, 45, 77, 141, 145–46

Drew, Jane, 30

Dubai, 61, 64, 76

dystopias: green capitalism and, 174; and utopias, 9, 15–16, 63, 174–76. *See also* utopias

Easterling, Keller, 185n3

ecocriticism, 20, 90, 145, 165–66; postcolonial, 19

ecologies: accessible and affordable urban spaces with minimal impact on, 4; impact of overconsumption and throwaway cultures on, 155, 173; informal settlements as sensitive to Kigali's ecologies and topography, 73–74; of infrastructure, 70–71; new market opportunities for investors from the planetary crises of economies and, 89. *See also* conservation; waste

economies: crashed postgenocide, 9; global extractive, 117; of recycling, 12, 17, 111, 122, 151; sustainability and green, 11, 54, 62; tourist-centric browsing, 152. *See also* green capitalism; popular economies; scarcity; trade

ecotourism, 27, 82, 151, 167. *See also* tourism

education, 53; Catholic mission, 34; facilities for higher, 77–78; public school teachers in, 130

electricity, 27, 44–48, 62, 81–82, 87–88, 96, 100, 119; infrastructure of, 102, 115, 123–26, 137; lack of, 103–4, 123; turning off the, 140. *See also* infrastructures

Electricity, Water, Sanitation Authority (EWSA), 124
employment, 97; artisanal brickmaking as important source of rural non-farm, 37; civil war and lack of, 47; "informal settlements" and availability of, 97; residents who live and work outside formal wage, 8; urban growth restricted by requiring all Rwandans to obtain official proof of formal, 44. *See also* labor
Engineers Without Borders, 69, 92, 95, 104
English, 69, 151, 187n2
Enlightenment project, 176
environmentalists, 20; activism of, 90; African artists as, 154, 165–66; dominant discourses of, 158; studies by, 191n2; Western, 179. *See also* activism; climate change; conservation
environmental justice, 158, 179, 188n3, 191n2. *See also* activism; environmentalists
Environmental Treatment Zones (ETZs), 70–71, 77, 138
Ernstson, Henrik, 77
ethnographies, 9, 19, 172, 179; of African cities, 90; realism of, 176; street, 20
ethnonationalism, 35, 47, 187n7
Europe, 147
evictees, 105, 111, 115, 119–20, 123–25, 135, 140, 152; people who engage in recycling as, 156, 165. *See also* demolitions; housing; markets
évolués, 33–34
extrastatecraft, 3, 85, 185n3

fashion, 24, 50; futuristic, 122; high-stakes labor of transforming clothing into, 153; near zero-waste, 12, 19, 147, 150, 154, 166, 171; as necessity and mode of creative self-making, 145; as nonverbal mode of communication, 154; recycling of, 24, 50, 126, 128, 142–57, 161–62, 164–65. See also *caguwa*; recycling; shoes; sneakers
floods, 3, 5, 9, 12, 16, 114; environmental hazards of disease and, 114; environmental injustice of, 41, 169; master plan-induced, 119, 169; wetlands for reducing erosion and, 70. *See also* wetlands
Fogler, Rob, 72
foreign direct investment, 9–14; housing market projections used to advertise Kigali as destination for, 100, 138; marketing of Kigali as competitive destination for, 60, 71, 84–85, 89; master plan projects built with, 152, 191n2; privatized sustainable urban futures built with, 12, 63, 104, 175; real estate and banking boom in Kigali attracting, 149. *See also* global finance; master plan projects
Formation Technologies, 72
fossil fuels, 11, 98. *See also* greenhouse gas emissions
France, 44
Fredericks, Rosalyn, 147, 156
French Entreprise Construction, 40
Friedman, Milton, 67
Fry, Maxwell, 30

Gastrow, Claudia, 103
Gaugler, Jennifer, 37
genocidaires, 52
genocide, xv, 4, 7–8, 14–18, 25, 38, 57, 63, 71–72, 145, 185n4; conditions in place for the, 47–48, 187n7; neighborhoods repaired after the, 115; survivors of the, 47, 49, 51. *See also* Hutu; Tutsi; violence
geographies, 40, 66; Black, 178
Germany, 29–30
Ghertner, Asher, 61
Ghosh, Amitav, 19, 21–22, 176
Gibson, William, xv; Bridge Trilogy, 116
Gidwani, Vinay, 146
global finance, 3; agendas of hedge funds of, 60; and tourism, 53, 65. *See also* foreign direct investment
global financial meltdown (2007–8), 10
Global North, 13
Global South, 8, 13; cyberpunk spaces in the, 116; development issues in the, 69; sites of disposal in the, 146; urban crisis in the, 19
Goodfellow, Tom, 51–52, 122–23
Gordillo, Gastón, 146
Gotham, Kevin, 67
Great Lakes Energy, 104
Greenberg, Miriam, 11, 13, 67, 188n3
green capitalism, xv, 3–5, 24, 57, 141; African alternatives to, 7, 25, 112–41, 174; corporate logics of, 178; destruction brought on by, 7, 173; and dystopias, 174; planning fictions that drive, 88–89, 112; projects of, 91–92, 112, 167; sustainable urbanism as, 10–12, 21, 62–63, 144–45, 171; synergistic promises of, 176. *See also* capitalism; economies; green commodities; master plan projects; sustainability
green commodities, 4; manufacture of effective consumer demand for, 115; new markets for,

24, 89; as solutions for African cities in crisis, 112. *See also* green capitalism
greenhouse gas emissions, 98, 190n1.
 See also climate change; fossil fuels
gukatakata, 127
Günel, Gökçe, 11
Guyer, Jane, 127

Habyarimana, Juvénal, 44–45, 47–48
Hamitic hypothesis, 29–30
Hansen, Karen Tranberg, 145, 153
Harare, 127, 175. *See also* Zimbabwe
Harvey, David, 66
health care, 62; private health clinics for, 113; universal, 175
Hecht, Gabrielle, 177
Hickcox, Abby, 65
hip-hop scene, xi, 165, 179. *See also* media
Hoffman, Danny, 177
Hollywood, 174
Hong Kong: Kowloon Walled City, 116
Hopkinson, Nalo, 118
Hotel Rwanda (film), 72
housing: affordable, 6, 12, 41, 46, 49, 53, 74, 77–78, 88–95, 100, 106–9, 112, 133, 138–39, 171; African "problems" of infrastructure and, 12, 95; Belgian missions as engaged in experiments with brick, 37; building boom in the late 1990s in Kigali, 49–50; foreign-owned estates of, 122; government twenty-year plan to accommodate commercial sectors and new, 46–47; late nineteenth-century Rwandan, 35; off-grid technology for, 104; politics of infrastructure and, 28, 73–74; reconstruction of nineteenth-century Rwandan reed, 36*fig.*; shortage of, 4, 50, 74, 88–89, 91–93, 106; sustainable solutions to fictional crises of inadequate, 88–90, 100–101, 103, 106, 112; unauthorized construction of, 26–28, 82–83. *See also* architectures; Batsinda estate; demolitions; evictees; informal settlements; infrastructures; land
Huchu, Tendai: "The Sale" (short story), 22, 62–63, 67, 78, 174–77
Hudani, Shakirah: *Master Plans and Minor Acts*, 18, 51, 67, 134, 161, 186n4, 189n4, 190n8
human rights, 130, 171
Human Rights Watch, 130, 187n10
Hutu, 29, 35, 39, 44, 72. *See also* genocide

Ibadan, 17
ibibanza, 148–49
ibimina, 128–29, 135, 150, 189n10
ibitenge, 162
ibiti na ibyumba, 38, 98
ibyagombwa by'ubutaka, 82
Igihe (Kinyarwanda online daily), 159–60
igishushanyo, 68
igisope, 6
Iheka, Cajatan: *African Ecomedia*, 21, 90, 97, 153–56, 173–74, 179
imibereho, 41
imidugudu, 5
imiyenda igezweho, 154
imperuka, 5, 9
Indians, 35
indigenous peoples, 34, 37; nature conservation as justification for expulsion and eviction of, 65; nineteenth-century displacement of, 187n1
inequalities, 7; social and economic, 18, 117; sustaining capitalism through, 130, 132. *See also* injustices
informal settlements, 23, 46, 50, 90; cheap cost of the redevelopment of, 81; language of "informality" as applied to, 105, 107, 138, 171; range of definitions of, 93–94, 185n5; as sensitive to Kigali's ecologies and topography, 73–74; and sprawl, 73, 81; "threat" to Kigali's urban core of, 101, 115, 170. *See also* demolitions; housing; slums
infrastructures: African housing and, 12, 44, 93, 119; densely populated informal settlements with inadequate, 73, 88; green, 11, 70–71, 78, 104; hacking "state infrastructural power," 122–26, 133, 136–37; history of colonial and imperial, 77; investments in the maintenance and repair of urban, 66–67, 123–26; lack of drainage, 1, 3, 169–70; projects of urban, 89, 93, 137, 170; wetlands as important ecological, 70–71. *See also* electricity; housing; master plan projects; water
injustices: of displacement, 18, 86–88, 106; environmental, 41, 165, 170. *See also* inequalities
Interahamwe, 47–49
inzoga z'abagabo, 42, 43*fig.*, 50, 82, 123, 135–36, 187n5
Islam, 32–34, 46

Jackson, Steven, xi, 133
Jefremovas, Villia, 37–38, 41
Johannesburg, 12, 17, 117, 125. *See also* South Africa
Jones, Jeremy, 127

Kabiru, Cyrus, 153
Kabuye, Rose, 48
Kagabo, José, 33
Kagame, Paul, 9–10, 51, 60–61, 69, 85, 186n7
Kahui, Wanuri, 153
Kampala, 8, 30, 45, 77, 141, 146. *See also* Uganda
Kandt, Richard, 30, 32–35, 42, 46; *Caput Nili*, 29
karabasasu, 38–39, 41, 98, 102, 119
katziye, 22, 95
kavukire, 135
Kayibanda, Grégoire, 39–41, 44–45, 48
Kenya, 66, 153. *See also* Nairobi
Kigali, 2*map*, 31*map*; Bugesera airport of, 74, 77, 89; as capital of the African century, 7–9; Car Free Day in, 55–56; central business district of, 152; Containers (detention facility) in, 52; Duhahirane-Gisozi shopping complex in, 159–62; École Technique Muhazi of, 33; expanding state bureaucracy in, 79–80; founding of, 30, 32; Gasabo district of, 1; history of industrial capitalism absent from, 11; Hotel des Milles Collines in, 152; Kacyiru Hill of, 39; Kwa Kabuga in, 130–31; majority African residents of, 77; meeting, incentive, conference, and exhibition (MICE) district of, 55, 170; New City Center of, 74, 75*map*, 76, 76*fig*., 89; Nyarugenge Hill of, 1, 30, 33, 35, 36*fig*., 39, 42, 52, 77, 102, 119; Office of the Ombudsman of, 108; One Stop Center in, 80–84, 136; popular transportation networks of, 70; postgenocide population surge in the neighborhoods that surround the core of, 47, 148; postgenocide rebuilding in, 16–18, 25, 49–50, 96, 172; racialization through spatial hierarchies of, 28; RPF efforts to take urban control of, 51–52; steep slopes of, 23; urban authority in, 172–73. *See also* Kigali City Master Plan; Kigali Convention Center; markets; mountains; neighborhoods; Rwanda; wetlands
Kigali City Master Plan, xiv, 3–4, 7, 13, 19–25, 57, 60–63, 80, 85, 126, 133, 144–45, 159–62, 172; critics of the, 89; estrangement generated in the, 176–77; international awards of the, 85; marketing fictions of the, 84; reactionary green perfection in the, 21; tourist spaces and shopping malls predating the, 152; unauthorized construction and the, 27, 84, 118–26; vacant lots in the urban core left behind by the implementation of the, 159–62, 164, 167. *See also* consultants; Kigali; master plan projects

Kigali Conceptual Master Plan, 61–63, 68–70, 73, 91–95, 100; consultants working on the, 92–93; prototype Batsinda house as presented in the, 103*fig*.
Kigali Convention Center, 55, 138, 152. *See also* Kigali
Kigali Economic Development Strategy (KEDS), 9
Kigali Institute of Science and Technology Studies (KIST), 9
Kigali Land Use Master Plan, 80. *See also* master plan projects
Kigali Today (online newspaper), 164
Kimari, Wangui, 77
Kimicanga, xiv, 1, 3–4, 59*fig*.; affordable housing in, 53; as built through historically established processes of land transfers and clay architectures, 46; conventions of city building established in, 115; as cultural center, 6; demolitions in, 1, 4, 6–7, 12, 17, 56, 85, 110, 120, 159, 167, 169, 185n6; electricity and drainage infrastructure installed in, 44; as flood prone, 6; former residents of, 111–12, 123; landowners in, 42, 49; large sand deposits across the wetlands in, 98; rich deposits of clay in, 41. *See also* neighborhoods
Kinshasha, 12, 17, 45, 127, 152. *See also* Congo
Kinyanjui, Mary Njeri, 129
Kinyarwanda, xi, 4, 22, 118, 122, 143, 154, 156–57
Kiswahili, xi, 34, 154
Kiyovu, xiv, 1, 39–42; cy'abakire/cy'abazungu, 40, 78, 186n4; elite enclave of, 34, 55, 96. *See also* Kiyovu cy'abakene; neighborhoods
Kiyovu cy'abakene, 17, 40–58, 57*fig*., 78–79, 81, 83–107; building of, 186n4; conventions of city building established in, 115; demolitions in, xiii, 58, 78–88, 92, 96, 101, 106–12, 119–20, 133, 137, 159, 169, 171–72; displacement of residents of, 106–7, 111; former residents of, 111–12, 123; near zero-carbon technology used to build, 105. *See also* Kiyovu; neighborhoods
Klein, Naomi, 67
knowledge: dispossession of, 104–6; plunder of African technology and, 105, 112; of sustainable urbanism, 5; wide distribution of construction, engineering, and legal, 17
kubaka kugikoboyi, 133–34, 136–37
kuchagua, 143
kugikoboyi, 5
kukiya-kiya, 127

Ku Klux Klan, 64–65
kwinjera, 122
Kyoto Protocol, 10

labor: corvée obligations of, 34–35; expert, 4; as forced and supplied by land chiefs under Belgian colonial rule, 32–33; precarious immigrant, 147. *See also* employment
Lagos, 12, 158. *See also* Nigeria
Lake Kivu, 32
Lake Victoria, 32
Lamarque, Hugh, 121
land, 41–46; alternative titling schemes relating to, 42, 44, 46; architectures and processes of transaction relating to, 45, 49; bureaucracy of offices relating to, 42; new system of the tenure of, 80; promises of reform of, 46; regimes of the colonial and postcolonial tenure of, 80. *See also* housing; land ownership
land ownership: affordable rental housing market in the hands of Rwandan, 100; property documents required to show, 94. *See also* land
law: housing, 46; land tenure, 41; master plan projects as, 139; national expropriation, 139. *See also* master plan projects
Le Guin, Ursula, 174
Lomé, 17
Longman, Timothy, 61
Los Angeles, 69
Louis, William, 30
Luanda, 8

magazini, 148, 154
Mail & Guardian (South African newspaper), 8
Makhulu, Anne-Maria, 45
Manirakiza, Vincent, 50, 137
Maputo, 28, 45. *See also* Mozambique
markets: Biryogo Market, 142–45, 143*fig.*, 162, 163*fig.*, 164–65, 190n8; Kimisagara Market, 26, 143–44, 164; Nyabugogo Market, 49–50, 56, 144–53, 144*fig.*, 154, 159–62, 164, 164*fig.*, 167, 190n8; Nyamirambo Market, 144, 164; Nyarugenge Market, 53, 132, 148, 162, 189n14. *See also* evictees; Kigali; Nyabugogo
master plan projects, 18–19, 125–26, 145; Central Business District Sub Area Master Plan (2010), 150; codes and regulations involved in, 134–36, 170; destruction of neighborhoods involved in, xiii, 78–79, 122–23, 133, 138, 140, 167, 169, 189n4; discarded printout to cover up an X of one of the, 135*fig.*; infrastructural developments as part of, 77, 137, 170; malfunction of, 145–46, 167; multistory shopping complexes promoted as, 152; theory of twentieth-century modernist, 61. *See also* construction; demolitions; foreign direct investment; green capitalism; infrastructures; Kigali City Master Plan; Kigali Land Use Master Plan; law
McKittrick, Katherine, 65, 178
media, 4–5, 13; Africanfuturist, 22, 155; Eurocentric practices of, 97; feature pieces in international, 84; high-carbon corporate production of, 21; images of undignified Africans in the global, 155; imperfect African, 21, 97, 173–74, 179; Kinyarwanda, 4, 6, 116, 119, 122, 131–32, 139, 159, 161, 164, 185n2; as made with caguwa, 145, 154–57; perfect, 174; promotional reporting and public releases in the, 72, 116; Rwandan genocide in the, 14; social, 20, 80, 152; sustainable plans as, 20, 71–73. *See also* hip-hop scene; online journalism
Meintjes, Louise, 97
merchants: Baluchi, Omani, and Afro-Arab, 33; Omani and South Asian, 33. *See also* trade
migrant workers, 39–40, 147. *See also* migration
migration: increase of rural-to-urban, 48, 189n4; restriction of rural-to-urban, 39. *See also* migrant workers
mining, 166
missionaries, 35; Belgian, 37
Mobutu Sese Seko, 48
Monrovia, 17, 177
Monteiro, Fabrice, 153; *The Prophecy* (Africanfuturist photo essay), 155–56, 176–77
More, Thomas: *Utopia*, 15
Morton, David, 28, 45
mountains, 26. *See also* Kigali
Mount Jali, 79
Mount Kigali, 26–27, 37, 53, 79, 82, 140–41
Mouvement Révolutionnaire National pour le Développement (MRND), 44
Moyo, Dambisa: *Dead Aid*, 10
Mozambique, 45. *See also* Maputo
mucaro, 49
mumanegeka, 114
Mumbai, 173
mumujyi, 49
Murri-Freres, 40
Museveni, Yoweri, 45, 47
Mususa, Patience, 17

Mutongi, Kenda: *Matatu*, 20
mwami, 30, 32, 35, 46
Myers, Garth, 7, 20, 62, 106

Nairobi, 1, 12, 30, 77, 94, 125, 129. *See also* Kenya
National Geographic Travelogue, 55, 63
neighborhoods: Banya/bitmahe, 5; Biryogo, 33–34, 42, 157; Busanza, 139, 141, 171; Commercial, 126; Cyahafi, 42, 102, 102*fig.*; Dobandi, 5, 119–20, 126; Dusheni, 111*fig.*, 112, 119–21, 126, 135; Gacuriro, xiv, 56, 58*fig.*, 85, 111, 120, 123; Gikondo, 140–41, 155, 187n11; Gisozi, 159; Gitega, 42; Kagugu, 128, 140–41; Kazaire, 17, 56; Kibagabaga, 113; Kimihurura, 39;; Kosovo, xiv, 1, 3–5, 119–20, 169–70; Lafreshaire, xiv, 56, 133, 137; Marato, 126, 148–49; Mashirahame, 149; Mateus, 126, 147; Muhima, 49, 81; Mukidelenka, 114; N'Djamena, 133; Nyamirambo, 26, 33–34, 39, 44, 104, 115, 155; Nyarutarama, 113, 152; Sodoma, 119–20, 126; Tapis Rouge, 33–34, 42. *See also* Bannyahe; Kigali; Kimicanga; Kiyovu; Kiyovu cy'abakene; Nyabugogo
neoliberalism, 17, 67, 122. *See also* capitalism
Neptune Frost (Kinyarwanda/Kirundi cyberpunk musical film), 22, 116–17, 122, 166, 176–78
Netherlands, 80
New Orleans, 67–68
New Times (online journal), 151
New York City, 67; New York Stock Exchange (NYSE) in, 186n11
Nigeria, 153. *See also* Lagos
Nollywood, 21, 97, 173
non-government organizations (NGOs), 92
Nyagenzi, Masabo: "Kavukire" (pop song), 186n3
Nyanza, 30, 35
Nyabugogo, 49, 82, 126, 131, 133, 137, 145, 147–53; "City Bazaar" planned for, 150–53, 151*fig.*, 159; demolition of, 154, 158; Kwa Mutangana in, 164; shopping complex planned for, 151–53; two sustainable futures of, 150–53. *See also* markets; neighborhoods

Oh, Sehoon, 85
oil, 61
Okorafor, Nnedi, 21
Omani, 35
Omelsky, Mathew, 118
online journalism, 4. *See also* media
OZ Architecture, 10, 13–14, 28, 53–56, 61, 67–81, 84–85, 88–89, 92, 95, 150; master plan of, xiii, 100, 102, 171; vision for the location of the New City Center in relation to other locations in Kigali of, 74, 75*map*, 76, 83; vision of the New City Center of, 76, 76*fig.* *See also* consultants

Paris, 61
Pentecostal churches, 114
Piot, Charles, 32
plastic bag ban, 84
politics: of infrastructure and housing, 28, 73–74; of repair, 16–19, 28, 51, 121–22; street, 129–33
Polly, Jay, 149, 190n5; "Caguwa" (music video), 154–58, 190n6
popular culture. *See* media
popular economies: of disaggregation and discounting, 114, 126, 128–30, 133; rebuilding of, 19, 50–53; recycling through networks of locally owned businesses and, 70–71; and spaces, 23–24; street, 4–5, 50–53, 96, 119–21, 126–35, 137, 146, 149, 178; and technologies, 5, 24, 89; zigzagging as the ability to manipulate and navigate the illicit street, 127. *See also* economies; recycling; repair
postapocalypse, 8; utopia and, 14–16
poverty, 87; affordable housing for those in, 93; causal processes of scarcity and, 21; conventional approach to housing scarcity and, 93; and proximity to the wetlands, 114; "world class" urbanism and urban, 89
public health, 41, 97

Quayson, Ato: *Oxford Street, Accra*, 20, 127

race: contemporary development projects and, 77; European constructions of, 32, 34; legal and material infrastructures of colonialism as codifying and producing identities of, 34. *See also* racism
racism: anti-Black, 65–66; environmental, 21, 153, 157; in industries that claim to disavow it, 105. *See also* race; segregation
railways, 32
recycling, 142–67; art of, 153–58; as conversion of material waste into value, 144, 146–47, 153, 156, 166–67; and discounting, 20, 53; economies of, 12, 17, 111, 122, 151; of fashion, 24, 50, 126, 128, 142–57, 161–62, 164–65; low-budget and low-fi popular cultures of, 174; networks of locally owned businesses and popular economies for, 70–71; vital labor

of repair, maintenance, and, 18, 126–33, 145–47; wasted spaces converted into value through, 145–47, 160–62. *See also* fashion; popular economies; repair
refugees, 14, 117, 187n8; in Burundi, 48–49; in Congo, 48–49; in Uganda, 48–49
repair: building technologies of, 112, 117, 139–40; Kigali as center of postgenocide, 28, 39, 48–51, 147, 167, 172; and maintenance as used to support and legitimate technocratic authority, 118; politics of, 16–19, 28, 51, 121–22; of popular city from sustainability-induced catastrophe, 90, 133, 146; popular processes of construction and, 18, 25, 50–53, 68, 72, 117–22, 139–41; as rendered impossible by state domination and a culture of compliance, 18; unauthorized, 22, 24, 27–28, 109–12, 115–22, 126–33, 137–40, 167, 177. *See also* built environments; construction; popular economies; recycling
repair studies, 117–18
reparations, 86–90, 106–7, 109, 186n4. *See also* displacement
research facilities, 77
Rhodes, Cecil, 32
Robinson, Jennifer: "Global and World Cities: A View from Off the Map," 186n9
rubanda rugufi, 68
ruswa, 136
Rutarindwa, Mwami, 29
Rwabugiri, Mwami, 28–29
Rwanda: artisanal brickmaking in rural, 186n2; cell (neighborhood-level) offices staffed by ruling-party secretaries of the government of, 42, 44, 46; civil war in, 47–48, 147; community policing in, 121; currency crash in, 17, 147–48, 187n7; formal independence of, 28; German colonial presence in, 29; Ministry of Gender of, 131–32; Ministry of Health of, 115; Ministry of Infrastructure of, 92, 121; Ministry of Trade and Industry of, 159–60, 162; national census (2001) of, 94; national pension fund (Caisse Sociale) of, 106–10; oral traditions of, 28; as placed under Belgian "tutelage" by a League of Nations mandate, 32. *See also* Kigali
Rwanda-Congo border, 40, 48, 96
Rwanda Cooperative Agency (RCA), 148
Rwanda Defence Force, 114, 131
Rwanda Housing Bank, 88
Rwanda National Police, 52, 108, 130, 161

Rwandan Development Board (RDB), 84–85
Rwandan Patriotic Army (RPA), 47–49, 187n8
Rwandan Patriotic Front (RPF), 47–48, 51–53, 85, 96, 115, 123, 172

Sachs, Jeffrey, 56
sanitation, 53; cooperatives of, 130; infrastructure for, 114, 138. *See also* sewage
scarcity: construction of, 101–4; conversion into environmentally progressive solutions of, 112; crisis of housing, 100, 106; design of, 92–93; and ecological destruction wrought by global throwaway cultures, 158; economies of, 97; production of, 3–4, 88–92, 106–9, 112; projection of, 93–95; as source of consumer demand rather than as a barrier to economic growth, 93. *See also* economies
science: nineteenth-century European race, 29; pseudo race, 44
science fictions, 21–22; African, 174; colonial frontiers and, 28–30; histories of colonial-era, 23; postcolonial, 118, 176; in time of climate crises, 176
Scott, James, 61
Secretariat des Missions d'Urbanism et d'Habitat (SMUH), 39
segregation, 33–34, 46, 77. *See also* racism
Senegal, 153, 155
sewage, 71; absence of a citywide central system of, 114. *See also* sanitation; water
sexual harassment, 136
sex workers, 52, 96
Shanghai, 61
shoes, 149–50, 154. *See also* fashion
Simone, AbdouMaliq, 117
Simpson, W. J., 30
Singapore, 3, 7, 12, 61, 64, 72, 76
Sirven, Pierre, 44
Situationists, 121
Slum Drip: "Nituebue/Nitwebwe" (music video), 164–65
slums, 6–7, 23–24, 46, 50, 73, 89–90; purchasing power of dwellers in, 91; range of definitions of, 93–94, 185n5. *See also* informal settlements
sneakers, 149, 165. *See also* fashion
social safety nets, 66–67
solidarity entrepreneurialism, 129
Sommers, Marc: *Stuck*, 134
sorghum, 114
South Africa, 45, 66, 173–74, 177. *See also* apartheid; Cape Town; Johannesburg

spaces: popular economies and, 23–24; popular urban, 178; recycling of, 161–66; tourist, 152; wasted, 145–47, 160–61, 165, 167. *See also* urbanism
Standard Gauge Railway, 32
structural adjustment programs, 63
subúrbios, 45
sugarcane, 114
Surbana, 79–85, 89, 98, 150, 186n8, 191n3; marketing campaign of, 170–71; single-use zoning codes of, 169. *See also* consultants
surveillance, 121, 123, 133–36; hacking into systems of, 141
sustainability, 10–12, 54, 57–58, 61–62; delayed, failed, and abandoned efforts to achieve, 144, 159; as "green veneer," 70; myth of, 191n2; perspective of the design of, 104; requirements of textbook, 100; solution to the shelter crisis as model of, 95, 101. *See also* green capitalism; urbanism
sustainable commodities, 91
Suvin, Darko, 16
Swahili, 32–35, 42

Tanzania, 32, 48. *See also* Dar es Salaam
Tema, 17
tourism, 3, 13, 113; high-end conference, 56, 77, 126; Kigali as center of global finance and, 53, 158, 170; Nyabugogo Market as transformed into destination for, 150, 161. *See also* ecotourism
trade: global, 28; international leather and ivory, 32; late twentieth-century liberalization of, 147; regional caravan, 33; in secondhand clothing imports, 145–48. *See also* economies; merchants
traffic: facilitating the flow of, 151; jams of, 148; protection of pedestrians from automobile, 159
transportation, 97; infrastructure of, 66; master plan for, 80–81; options for, 109; popular, 6, 70
Trouillot, Michel-Rolph, 4, 7, 106
Tsamaase, Tlotlo: "Behind Our Irises" (short horror story), 22, 90, 104–6, 176–77
Tutsi, 4, 7–8, 14, 16, 29, 35, 44, 47–48, 57, 63, 72, 145, 187n8. *See also* genocide
Twa, 29, 35

ubucuruzi, 51
ubusa, 146, 164
ubwrira kabiri, 28
Uganda, 47–49; Kigali residents in refugee camps in, 48–49. *See also* Kampala

Ugandans, 4, 40
umucanga, 41
umudugudu authorities, 27, 51, 83, 111, 188n5. *See also umuyobozi w'umudugudu*
umuganda, 5, 26–27, 116, 124, 126
umukoboyi, 134
umunani, 41
Umuseke (Kinyarwanda online newspaper), 161–62
umushumba, 134
umutarage, 149
umutekano, 115
umuyobozi w'umudugudu, 26
umuzunguzayi, 51, 128, 131, 133, 136
United Kingdom: Department of Foreign International Development of the, 80; plan for the removal of political asylum seekers from the, 171
United Nations, 3, 7, 11, 50; commodity trade database of the, 189n2; definition of sustainable urbanism of the, 69; Media project Integrated Regional Information Networks (IRIN) report of the, 187n10; Millennium and Sustainable Development Goals rankings of the, 56; Sustainable Development Goals of the, 90
United Nations Environment Programme, 167
United Nations-Habitat, 7, 20, 23, 70, 74, 91, 101, 110; Scroll of Honor Award (2008), 85, 106; *Slums of the World* (report, 2003), 73, 93–95; statistics on housing deficits of, 85, 88, 90, 94–95
United Nations Millennium Development Goals, 51, 70
United States, 3, 7, 12, 44, 147; plan for the removal of political asylum seekers from the, 171; Wall Street in the, 8
Urban Boyz: "Caguwa" (music video), 154–58, 190n6
urban humanities, 19–22, 174, 179
urbanism: critique of top-down, 24; historical significance and everyday politics of, 44–45; investment-driven, 13, 63; "laissez-faire," 45; popular, 3, 178; privatized, 22, 62, 66–67, 79, 85, 175–76; punk, 22, 113–41, 171, 189n6; sustainable, xv, 3–12, 15–23, 56, 62–64, 79, 85, 88–89, 97, 112, 116, 133–40, 145–46, 161, 166–67, 170–74; zero-tolerance, 175–76. *See also* African cities; capitalism; spaces; sustainability; urban planning
urban planning, 5–6, 14–17, 136; as case study of postgenocide national state building, 28; colonial and independence-era modernist

projects of, 17; definition of informality in, 98; dispossession brought about by, 17; as instruments that go beyond the global image, 60; twentieth-century modernist, 15, 66; and urban authority, 144; visual aesthetics of, 12. *See also* architectures; brand image; built environments; design firms; urbanism; utopias

Usumbura (Bujumbura), 33

utopias, 3, 13; dystopias and, 9, 15–16, 63, 174–76; as genre, 15–16; green capitalism and, 4, 55–56, 174; investment opportunity as, 24, 72; and postapocalypse, 14–16, 56; in urban planning, 16, 63, 89–90, 118, 133. *See also* dystopias; urban planning

Uzeyman, Anisia, 117–18, 166, 179

violence: discard industries and slow, 155; of dispossession, xv, 17, 112; extraeconomic, 108, 185n6; of extralegal detainment and street raids, 130–33; postconflict recovery and mass, 8, 14; state, 90–91, 117–18, 131, 178. *See also* genocide

Vision 2020 plan, 51

Vumbi, Danny: "Ni Danger" (pop song), 157–58

Wakanda, 61

waste: conversion into value of, 71; fashion industries with low, 12, 19, 147, 150, 154, 166, 171; streams of, 11, 147, 155. *See also* conservation; ecologies

water, 44–48, 62, 69, 81–82, 87–88, 100, 119; access to, 94, 96; infrastructure of, 91, 102, 115, 123–26, 137; lack of, 103–4, 123; turning off of, 140. *See also* infrastructures; sewage

wetlands, 1, 4, 6, 23, 26; as colonial-era barrier to the center of town in efforts to prevent Rwandans from entering Kigali, 41; demolition of everything between the original core and the, 106; hazards of living near the, 115, 170; Kigali as built with earth, clay, and sand secreted by the ecosystems of the, 28, 35, 37, 39–41, 44, 97; Kigali directly before independence as small collection of segregated camps on top of Nyarugenge Hill surrounded by, 33, 40; poverty and proximity to the, 114; unregulated urban growth as threat to the, 89; and waterborne diseases, 97, 114; water obtained from the, 103. *See also* floods; Kigali

Williams, Saul, 117–18, 166, 178–79, 189n7

women, 129–31; running of discount street economies by, 130–31, 146, 178

World Bank, 7, 40, 44, 91; "Ease of Doing Business" rankings of the, 56, 85

World Economic Forum on Africa, 131

World War I, 32, 35, 64

World War II, 64

Worthington, Carl, xiv, 9–10, 13, 16, 60–61, 64–66, 69, 72, 76, 85

YouTube videos, 116. *See also* media

Yuhi V Musinga, Mwami, 29

Zaire. *See* Congo

Zambia, 17, 145

Zanzibar, 30

Zimbabwe, 62, 127. *See also* Harare

zoning, 53; and building codes, 8, 83, 100, 121, 134; digital databases for, 60, 79–82, 84, 135; land use and policies of, 65. *See also* building codes

Founded in 1893,
UNIVERSITY OF CALIFORNIA PRESS
publishes bold, progressive books and journals
on topics in the arts, humanities, social sciences,
and natural sciences—with a focus on social
justice issues—that inspire thought and action
among readers worldwide.

The UC PRESS FOUNDATION
raises funds to uphold the press's vital role
as an independent, nonprofit publisher, and
receives philanthropic support from a wide
range of individuals and institutions—and from
committed readers like you. To learn more, visit
ucpress.edu/supportus.